DON'T TRUST
DON'T FEAR
DON'T BEG

100 Days as a Prisoner of Putin
– The Story of the Arctic 30

BEN STEWART

First published in 2015
by Guardian Books, Kings Place, 90 York Way, London, N1 9GU
and Faber & Faber Ltd, Bloomsbury House,
74–77 Great Russell Street, London, WC1B 3DA

This paperback edition published in 2016

Text and plates designed by seagulls.net
Printed and bound by CPI Group (UK) Ltd, Croydon CR0 4YY

A CIP record for this book is available from the British Library

ISBN 978-1-78335-078-0

2 4 6 8 10 9 7 5 3 1

'Dear future generations: Please accept our apologies.
We were rolling drunk on petroleum.'

Kurt Vonnegut

The author's royalties from sales of this book are being donated to groups that fight for environmental protection and the rights of political prisoners: SaveTheArctic.org, PlatformLondon.org, Agora-Sofia (www.openinform.ru) and Human Corpus (www.corpusmedia.org).

For the families of the Arctic 30, and for my own

NOVAYA ZEMLYA

BARENTS SEA

Murmansk

Arctic Circle

FINLAND

RUSSIA

Greenpeace ship *Arctic Sunrise* boarded near the Prirazlomnaya rig

St Petersburg

200 miles

RUSSIA

FOREWORD

Hi, Paul McCartney here.

1968. That was quite a year. The people were on the streets, revolution was in the air, we released the White Album, and perhaps the most influential photograph of all time was taken by an astronaut called William Anders. It was Christmas Eve. Anders, his navigator Jim Lovell and their mission commander Frank Borman had just become the only living beings since the dawn of time to orbit the moon. Then, through the tiny window of their Apollo 8 spacecraft, their eyes fell upon something nobody had seen before, something so familiar and yet so alien, something breathtaking in its beauty and fragility. 'Oh my God,' Anders cried. 'Look at that picture over there! There's the Earth coming up. Wow, is that pretty!'

'You got a colour film?' he asked the others. 'Hand me that roll of colour quick, would you … ' For a minute or so, three human beings in a tin can nearly 400,000 kilometres from home scrambled furiously to fix a roll of film into their camera. Then Anders lifted it to the window and clicked the shutter and captured our delicate home planet rising slowly over the horizon of the moon. Earthrise. That single image made such an impact on the human psyche that it's credited with sparking the birth of the global environment move-ment – with changing the very way we think about ourselves. That was nearly half a century ago, the blink of an eye in the grand sweep of time, but something quite remarkable has happened since then. For as long as humans have inhabited the Earth, the Arctic Ocean has been capped by a sheet of sea ice the size of a continent. But in the decades since that photo was taken, satellites have been measuring a steady melting of that white blanket. Much of it has now gone, and it seems possible that for future generations the North Pole will be open water. Think about it. Since Earthrise was taken we've been so busy warming our world that it now looks different from space. By digging up fossil fuels and burning our ancient forests we've put so much carbon into the atmosphere that today's astronauts are looking at a different planet. And here's something that just baffles me. As the ice retreats, the oil giants are moving in. Instead of seeing the melting as a grave warning to humanity, they are eyeing the previously inaccessible oil beneath the seabed at the top of the world. They're exploiting the disappearance of the ice to drill for the very same fuel that caused the melting in the first place. That's why, in summer 2013, thirty men and women from eighteen countries sailed for a Russian Arctic oil platform, determined to focus global attention on the new Arctic oil rush. They saw how fossil fuels have come to domi-nate our lives on Earth, how the energy giants bestride our planet unchecked. They knew that at some time and in some place somebody had to say, 'No more.' For those activists that time was now and that place was the Arctic.

Their ship was seized, they were thrown in jail and faced fifteen years in prison. Millions of people from across the world raised their voices in support of the stand they took, including many from the great nation of Russia. The tale you are about to read is extraordinary. It is one of fear, hope, despair and humanity. But we still don't know how it ends. That is up to all of us. Including you. Please encourage your friends to help bring a hopeful conclusion to this moving story.

Paul McCartney, December 2014

INTRODUCTION

Frank Hewetson is lying on the upper bunk of a prison cell in the Russian Arctic, waiting impatiently for the effects of a Valium tablet to kick in. He's wearing woollen tights, two pairs of socks, three T-shirts, a pullover, a skull-gripping hat and earplugs. The hot incandescent bulb dangling from a wire above his head has just been switched off by the guards, and Murmansk SIZO-1 isolation jail is stirring.

He can hear boots stomping on the floor above his head, prisoners thumping the walls in cells down the corridor, the distant sound of screaming. Across the prison, windows are swinging open and ropes are being fed through bars, then lowered down the outside walls or swung from cell to cell.

Frank pulls a blanket up around his neck and holds himself against the cold biting air. He is forty-eight years old, he has a wife and two children back in London and he's charged by the Russian state with piracy – a crime that carries a minimum sentence of ten years in a country where 99 per cent of all trials end in a verdict of guilty.[1]

He opens his eyes into narrow slits and looks down. One of his cellmates, Boris, is bent at the waist and pressing his ear against the plughole of the sink, an expression of strained concentration on his face. Boris is a short man with olive skin, muscles like marble, a permanent wrap of stubble on his face and a forehead so narrow that his hairline nearly merges with his eyebrows.

He's charged with double manslaughter.

Frank's other cellmate, Yuri (multiple counts of assault by Taser), is feeding a rope out of the window and whistling to himself. He's younger than Boris, not much meat on him, sallow skin and greasy black hair. Minutes from now this rope network, known as the *doroga* – 'the road' – will connect almost every cell along the outside walls of the jail, allowing the prisoners to communicate with each other and share contraband. It is a physical internet through which power is projected and justice dispensed by the mafia bosses who control much of this place.

With relief, Frank senses his mind becoming foggy. The air no longer stings his cheeks and he can't feel the wire mesh digging into his back through the thin mattress. Thank Christ for those drugs. Every night when the prison awakes the pills allow him to slip into something approaching sleep. He secured the Valium prescription five weeks ago after experiencing what the authorities thought was a cardiac arrest but which was, in reality, a panic attack brought on by the prospect of spending ten to fifteen years in a Russian jail. He was sped to hospital and bundled into a wheelchair then pushed through the corridors at breakneck speed by an armed guard. Patients and doctors dove into doorways to avoid being run down as Frank careered towards an emergency consultation, wires trailing from electrodes stuck to his bare chest, the guard singing lines to himself from the back catalogue of Depeche Mode.

Boris stands up straight and looks at Frank quizzically. 'Frank,' he hisses. 'Come come come. Frank!'

Frank closes his eyes, pretending to sleep, but a moment later he can feel Boris's breath on his face. It smells of potatoes and fish-head soup.

'Fraaaank. Come come.'

'Boris, piss off and leave me alone, all right.'

'Come, Frank. Come.'

He's pointing towards the sink. Something in his voice is utterly, irresistibly insistent.

'*Frank!*'

'Jesus, Boris. What?'

'Come!'

Frank rubs his eyes, pulls out the earplugs, swings his legs over the edge of the bunk and grudgingly jumps to the ground. Boris slaps him on the back then leads him over to the sink. Yuri ties off the rope, crosses the cell, kneels down under the sink and starts unscrewing the U-bend. Boris kneels down next to him and together the two Russians strain hard, pulling the pipe away from the wall until – with a scraping metallic *pop* – it comes clear.

'Frank, sit.'

Frank scratches his head. The air is filled with thumping and banging as the rope network comes alive. Soon the prisoners will be using it to share illicit letters, sugar, mobile telephones, an underground satirical newspaper and perfumed cigarettes given as gifts by prisoners to lovers they have never met and never will.

His cellmates are staring up at him with imploring eyes. Boris is clutching the liberated U-bend like it's a glass of beer.

7

Slowly, hesitantly, Frank lowers himself to the ground then Boris pushes Frank's head down, at the same time twisting the U-bend until it's pressed against Frank's ear. Frank's eyes swivel in their sockets; he stares at Boris and he's about to say something when he hears a faint tinny voice.

'Allo? Dis is prisoner boss Andrey Artamov in cell four-one-zero. Is dat the Arctic firty?'

Frank gulps. 'Er … ' He hesitates then puts his mouth to the end of the tube. 'Yes, hello?'

'Is dat the Arctic firty?'

'Er … yes. Well, one of them.'

'I have friend of you here.'

'Right. Okay.'

Silence, then, 'Hello, Frank?'

'Yes?'

'Frank, this is Roman Dolgov, your Greenpeace compatriot from the cell above you.'

'Er … hello, Roman. You seem to be somewhere in my U-bend system. How did you fit down there?'

'Ha ha, yes, this is funny, Frank. What you say is funny.'

'Roman, is this … are we talking on … is this a *telephone*?'

'This is prison telephone. I have to tell you, Frank, we have a problem.'

Roman is a 44-year-old campaigner from the Moscow office of Greenpeace, arrested with Frank and twenty-eight others when their ship was stormed by Russian commandos seven weeks ago. They'd held a protest at an Arctic oil platform operated by President Putin's state-run oil company, Gazprom, and now they're facing the full fury of the Kremlin.

'Roman, what's going on?'

'I speak with respected prisoners, Frank. They tell me you must talk to cell three-one-six. The cell opposite yours.'

'Okay. Why?'

'They say you must get the names of the Russians in that cell. They do not give their names, they do not go to *gulyat*' – the hour of exercise the prisoners are granted each day – 'and they have broken the *doroga*. They do not co-operate. The rope network on one wall is broken. Big problem.'

'Er … okay, Roman. So … so … I'm sorry, say again, *what* do they want me to do?'

'Francesco is also in their cell. You must ask him, what are the names of the Russians?'

Frank thinks for a moment. He rubs the fuzz on his head. His blond hair was closely cropped on the ship but now it's growing out. He hands the U-bend to Boris, stands up and opens a hatch in the door.

'Frankie!' he shouts.

In a door across the hallway a hatch opens and the face of 38-year-old Frenchman Francesco Pisanu – another of the Greenpeace detainees – appears.

'Yeah?'

'Francesco, what are the names of the Russians they've just put in your cell?'

'One moment.'

His face disappears. A minute later he returns.

'They will not tell me.'

'Francesco, you must find out the names of the Russians.'

'They will not tell me. They are scared to tell me.'

'Really?'

'They say they are scared.'

9

Frank kneels down, takes the U-bend and speaks into it. 'Roman, they won't say.'

'They will not say?'

'No.'

'Oh.'

At the other end of the pipe a conversation is conducted in Russian, before Roman returns.

'Okay, Frank. Good night.'

'That's it?'

'Good night, Frank.'

'Er … okay. Night, Roman.'

Frank leans back, still holding the pipe, tapping the end with a finger and biting his lip. Boris shrugs. Yuri grunts and pushes himself to his feet. Frank stares at the pipe for a moment before handing it back to Boris, then he stands up, sniffs, clambers back onto his bunk, pulls the blanket right up to his neck and lies there, staring at the ceiling.

An illegal telephone network fashioned from the prison plumbing system? Mafia bosses issuing orders through a U-bend? And this isn't even the strangest thing that's happened in the last two months.

'Christ,' Frank whispers to himself, shaking his head. 'How the fuck did I end up here?'

ONE

He lifts the binoculars, narrows his eyes and twists the dial to focus. His vision is flooded with blurry scarlet red. Frank turns the dial again and the view sharpens. He can see large white Russian letters, a helicopter deck protruding far over the water, the drilling tower standing out crisply against a blue sky.

He must have stared at that oil platform fifty times in the last twelve hours. It looks like a football stadium floating defiantly in the ocean, 180 miles north of the Arctic Circle. A half-million-tonne square block of metal and concrete with sheer red sides.[2] It's called the *Prirazlomnaya*.

Frank is standing at the bow of the Greenpeace ship *Arctic Sunrise*. Three miles of ocean separate him from the platform. He turns his head and the view through the binoculars fills with sweeping open water then the dark blue hull of another ship. It's a Russian coastguard vessel – the *Ladoga* – and it's slowly circling the platform, protecting it from protesters. Specifically from Frank Hewetson and his friends.

He sucks his teeth and lowers the binoculars. It won't be long now. Soon he'll know if his plan is good enough. Earlier today he launched a flotilla of RHIBs – inflatable speedboats – from the *Arctic Sunrise*. It was a dummy run to test the

Russians' reaction time. The coastguard took five minutes, maybe six, to launch their own boats. Frank watched them from the deck of the *Sunrise*. They were slower than his team. Slower than the Greenpeace crew.

We're ready, he thinks. It's going to happen. First light tomorrow.

There are two RHIBs on the Russian ship. Tomorrow morning he'll launch five from the *Sunrise*. He's got them beaten for numbers, but he'll need to surprise them too.

He lifts the binoculars and surveys the steel skeleton beneath the helicopter deck. That's where he needs to get the lines up. The team practised for days in a Norwegian fjord before they set sail. They constructed a fake helideck and attached it to the crow's nest of the *Arctic Sunrise* then bobbed in boats for hour after hour, firing ropes over it with catapults. After four days they were looping lines over the target nearly every time. But tomorrow morning they'll have to hit piping forty metres above their heads, with the Russian coastguard barrelling down on them in speedboats.

He turns the binoculars back to the *Ladoga* and blinks at a glint of brilliant reflected sunlight. He squints. Nearly three miles away a man in a blue blazer and peaked cap is standing at an open door, holding his own binoculars, watching him.

Frank Hewetson has been sailing with Greenpeace for two decades. He's been banned from the United States for crimes of moral turpitude, he's broken into seven polluting power stations in four countries, and he once blocked the take-off of a British Airways jetliner at Heathrow airport in a protest against climate change. Three years ago he was skewered by a grappling hook thrown by a French sailor while he was protesting against illegal bluefin tuna fishing. The hook passed cleanly through his left leg, then the Frenchman pulled on the rope, dragging Frank along the floor of a boat. Frank had to cut the rope with a knife to stop himself becoming the fisherman's latest illegal catch.

After twenty years leading direct action teams across the globe, he is the go-to guy if you want to scale an enormous piece of machinery being operated by a powerful company with a multi-million-dollar security operation. His colleagues call him 'The Colonel' – a nod to the confident cut-glass way he delivers orders, and because he chairs meetings wearing a World War Two tank driver's uniform.

He turns around and leans back against the railing. The bow of the *Arctic Sunrise* is dipping and rising gently. Frank lifts his baseball cap and rubs a hand over his scalp then he looks up at the bridge and sees three faces behind a broad wall of glass. One of them – a middle-aged man in a cream and blue sweater – has a pair of binoculars pressed against circular steel-rimmed spectacles. He has short black hair and a beard that's greying at the chin. His name is Dima Litvinov. His lips are moving but Frank can't hear what he's saying. If he could, he would hear an accent that sounds American but with a trace of something else. Dima is fifty-one years old, he was born in Russia and grew up in Siberia, where his family was exiled after his father challenged the Soviet regime. When he was twelve years old the Litvinovs were expelled from Russia and moved to New York.

Dima passes the binoculars to a young woman who lifts them to her face. She is Sini Saarela, an activist from Finland, thirty-one years old with a climber's lean physique, a bob of blonde hair with a centre parting and sharp blue eyes. Last summer she climbed the side of that oil platform – the one she's looking at now. She spent hours hanging from the *Prirazlomnaya* as she was blasted with freezing seawater fired from a cannon. The cold eventually forced her down, but a year has passed and now she's back for more.

Tomorrow morning her job will be to scale the platform again. She'll rig a pulley system, then the Greenpeace crew will lift a one-tonne, barrel-shaped survival pod equipped with state-of-the-art communications systems thirty metres above the water. There it will house three activists for as long as possible – days, maybe weeks – stopping the *Prirazlomnaya* from operating. That's the plan. But it will only work if Sini and her friends can reach the platform before the Russian authorities deploy their own RHIBs from the coastguard ship. They're in international waters, technically the Russians can't arrest them; last year they did nothing more than watch as Sini was drenched with Arctic water. But this time it feels different. The coastguard started tailing the Greenpeace crew a few hours after they sailed from Norway towards the platform. The Russians' radio messages were aggressive and uncompromising.

'*Arctic Sunrise*, *Arctic Sunrise*, under no circumstances will you approach the *Prirazlomnaya*. There is an exclusion zone of three nautical miles around the platform. You are ordered to stay far away from the *Prirazlomnaya*.'

Sini passes the binoculars to a man dressed in a white T-shirt and shorts. He's thickset with a handsome face and tanned skin. Pete Willcox is the 61-year-old American captain of the *Arctic Sunrise*. In three decades on Greenpeace ships he's tangled with commandos and coastguard officers more times than he cares to remember. He's sailed into nuclear test zones, swum in front of a US Navy destroyer and confronted Japanese whalers.

Frank first met him in 1991 when they plugged an outflow pipe at an Australian port where a mining company was pumping toxic effluent into the harbour. By then Pete Willcox was already a Greenpeace legend. Six years earlier he'd been the captain of the *Rainbow Warrior* when she was moored up in Auckland harbour, New Zealand. Pete was about to lead an expedition to protest against the French government's plan to detonate a nuclear weapon on the Pacific island of Moruroa. Just before midnight a limpet mine attached to the hull of his ship exploded. It had been laid by agents of the French secret service.[3]

The blast shook Pete awake. He thought his ship had been hit by another boat and he started racing through the *Warrior* checking on the crew, getting everyone out on deck. His friend, the Portuguese-born photographer Fernando Pereira, initially came outside but returned to his cabin to save his cameras. Minutes after the first blast, a second mine detonated. Fernando, a father of two young children, was drowned.

Twenty-eight years later Pete Willcox is leading another expedition, and another nation's security forces are determined to stop him. Before leaving Norway three days ago he sent his new wife a postcard. 'If the Russians keep their sense of humour,' he wrote, 'I think this is going to be a fun action.'[4]

Frank has known Pete Willcox and Dima Litvinov for most of his quarter century as a Greenpeace activist. But he met most of the *Sunrise* crew for the first time when they arrived in Norway last week. He watched them walking along the dockside with their bags slung over their shoulders – climbers, sailors and campaigners from eighteen different countries. The oldest was the captain, the youngest was Camila Speziale, a 21-year-old Argentine climber who quit her job as a receptionist to occupy a pod hanging from the helicopter deck of a Russian Arctic oil platform.

The *Sunrise* is fifty metres long, an icebreaker painted green with a riot of rainbow colours at the bow. When she sailed into that Norwegian fjord for four days of training, this was a ship of strangers. Now they're a tight crew. They spent the days firing catapults, climbing ropes, rigging the pulley system, lifting the pod. In the evenings they shared stories in the lounge. One night the ship's intercom exploded with two words.

'Northern lights!'

The crew ran out on deck and craned their necks. They draped their arms around each other's shoulders as a flag of transparent green fabric flapped slowly in the sky above their heads from one horizon to the other. The next morning they docked in the Norwegian port of Kirkenes. Then they sailed for the *Prirazlomnaya*.

Frank turns around and grips the railing. Across the water is the most controversial oil rig in the world. It's owned and operated by Gazprom, Russia's state-owned energy giant. Sometime in the next few weeks Gazprom will try to become the first company in history to pump oil from the icy waters of the Arctic. Until now the thick sea ice has made drilling here almost impossible, but as temperatures rise the oil companies are moving north, and if the *Prirazlomnaya* succeeds it will spark a new Arctic oil rush. That's why the *Sunrise* is here. That's why, right now, across this ship, thirty men and women are making final preparations to scale that platform and shut it down.

Frank leans over the bow and sees his reflection in the water. He breathes deeply and looks up. The last of the sun is sinking below the horizon. When it next appears, he'll give the order to go.

TWO

The portholes are screwed shut. The doors are closed. Nobody is allowed on deck. Not yet. To the coastguard officers defending the Russian oil platform three miles across the water, the *Arctic Sunrise* is sleeping.

But the Russians are wrong. It's 3 a.m. and every one of the crew of thirty is up and awake. Wide awake. Frank is pacing the hold, checking his watch. He's wearing a yellow drysuit under a life jacket and he's carrying a crash helmet with a transparent visor. Every few minutes he asks the British video journalist Kieron Bryan to join him at a porthole, where they lift the lid just a fraction and peek through, searching for sight of the sun, waiting for enough light to film the protest.

The sea isn't as flat as Frank had hoped it would be. He can hear waves slapping against the side of the ship, and when he looks over at the oil platform – lit up like a shopping centre – it sometimes disappears behind a swell of water.

Now the crew is making last-minute checks. Phil Ball, who will occupy the pod with the young Argentine Camila Speziale, is patting his chest, yanking karabiners, adjusting his helmet. *Have I got everything? Is it in the right place? Is it comfortable? Can I still grab hold of it if there's a water cannon firing in my face?*

At 3.30 a.m., through the porthole, Kieron sees the lip of the sun. Frank asks him if there's enough light to capture the action.

'I think so. Just.'

The crew is clustered together in teams, whispering to one another, checking the plan and checking again. 'Okay,' Frank announces. 'Everybody!' They look up, expectantly. A pause, then, 'We're doing it.'

Frank watches the activists blow out their cheeks and shake hands with each other. In front of him, Sini Saarela and Kruso Weber – a Swiss climber – are standing face to face, checking the other's kit one last time. Frank needs these two to perform today. If they can get up the side of that oil rig and hold their position, this thing might happen. He looks around. The activists are nervous, they're bouncing on their toes, their eyes are darting around the hold.

The boat drivers creep out onto the deck, using stairs and barrels for cover, thinking, as long as we stay low, as long as we can't see the coastguard ship, then they can't see us. Slowly, silently, the first RHIB – called *Hurricane* – is slipped into the water and moves up to the pilot door. Welshman Anthony Perrett helps the video journalist Kieron Bryan and the climber Kruso Weber to clamber in. Kieron presses 'record' and raises his camera, the black inflatable bow of the boat lifts and suddenly they're tearing around the *Sunrise* into open water. Ahead of them a spotlight breaks the dawn. The beam is coming from the coastguard ship *Ladoga* and within seconds it's sweeping across the rolling water towards them. Now the activists are bathed in blinding light, but they're still going full tilt, the boat is crunching through the waves. Already they can see the Russians launching their own boats.

A few seconds behind them a second Greenpeace RHIB – *Parker* – is rounding the bow of the *Arctic Sunrise*. In that boat are Frank and Sini. From the deck of the *Sunrise*, the pod – white and blue, built specifically for this moment – is being lowered into the water. Watching through binoculars from the bridge of the *Sunrise* is Dima Litvinov. He lifts a radio to his mouth and barks, '*Prirazlomnaya, Prirazlomnaya,* this is *Arctic Sunrise.*'

There's a crackle of static, then, '*Arctic Sunrise,* this is *Prirazlomnaya.*'

'This is a peaceful action, a non-violent protest against oil drilling and the threat that it represents to the Arctic environment and to the climate. There is no risk of damage to your property, we are in unarmed boats, we are not going to attempt to take over your platform. This is a peaceful protest. I repeat, this is a peaceful protest.'

An officer on the *Ladoga* breaks in. '*Arctic Sunrise,* halt all activity. Raise your boats!'

The Russian RHIBs are in the water now, but the activists' boats are already pulling up under the platform. It's huge, 120 metres long on each side. Anthony Perrett stands up in *Hurricane* and raises a catapult. It's more than a metre long, with a rubber sling that fires a lead shot attached to a bag of sand that pulls a thin line.

His first shot misses but his second shot arches over three metal bars then slowly slips down as a coastguard RHIB roars through the water towards them. The rope is four metres above his head, now three, he flicks it, it's nearly there. The Russian boat is close now, they can hear it rounding the corner of the platform. Anthony reaches up and grabs the line, attaches a thicker climbing rope to it and starts pulling on the other end, watching the rope rising higher and higher. He goes to pass it off to Kruso, it's a metre from the climber's hand, he's a second or two from clipping in and starting the climb when the coastguard RHIB tears around the side of the platform, white surf churning from its motor. It ploughs directly into their boat, then a masked Russian commando lunges at the rope with a knife and cuts it clean through.

Out in open water *Suzie Q* – the biggest of the campaigners' RHIBs – is towing the pod towards the platform with two smaller boats flanking her. But the pace of the flotilla is painfully slow. The boats are struggling through the water, it's like they're stuck in honey, and in the distance they can see a coastguard boat ramming a RHIB below the platform. Then suddenly – *thwack!* – *Suzie Q* lurches and a rope whips the water. The line has broken. Phil Ball looks back and stares at the pod, floating forlornly, pathetically unattached.

Silence, a static buzz, then Frank's voice on the radio. 'Dump the pod and get here. Now! All boats to the platform. All boats!'

The bow of *Suzie Q* lifts in the water and a moment later they're tearing towards the *Prirazlomnaya*. A few minutes later they've joined the action and, through a spray of water, Phil can clearly see two Russian boats carrying soldiers wearing black balaclavas over their faces, bodies camouflaged from top to bottom. Suddenly one of them pulls a knife and lunges at Kieron, trying to grab his camera, but the camera is attached to his chest by a cord. If the guy gets hold of it then Kieron's going to be pulled out of the boat. The Russian falls short, he leans down and stabs *Parker*'s rubber inflatable rim, then he reaches for his hip and pulls a gun. He points it at Kieron then swings it round so it's pointing at the chest of Italian activist Cristian D'Alessandro, who is standing at the bow of *Suzie Q*. The Russian is screaming something but nobody understands him. Cristian thrusts his arms into the air and shouts, 'Don't shoot, don't shoot!' A wave lifts both boats, their eyes are locked, the barrel of the gun is a metre from Cristian's chest, both of them are shouting at each other, then the wave dissipates and the boats fall and drift apart. *Suzie Q* throws up a wall of spray and pulls away with *Hurricane* just behind her, leaving the Russians in their wake.

Out at sea the pod is being retrieved by the *Sunrise*, but the activists are still determined to secure a climb team on the side of the platform. If they can get Sini and Kruso onto the rig they can unfurl banners and focus global attention on Gazprom's plan to drill for oil in the Arctic. They've lost the pod, but they can still make their stand.

On the eastern side of the platform *Parker* pulls up under a mooring line. Sini aims a catapult and fires a rope over it, checks it's secure, clips in and pulls. But a moment later a Russian boat speeds in, a coastguard officer pulls a knife and in a flash the rope above Sini is cut and she falls into the water. Her life vest inflates with the hissing sound of pressurised air. Frank reaches over the side of *Parker* and pulls Sini into the boat. She falls back, gasping.

On the other side of the platform *Hurricane* is pulling a huge sweep around the rig, with the chasing coastguard in her wake. Anthony spots another mooring line. He thinks he can get a rope over it. *Hurricane* pulls up, Anthony raises the catapult and fires a perfect shot. The rope is twenty metres up the side of the platform, Kruso grabs it and starts climbing.

Parker has abandoned the east side and a minute later is alongside *Hurricane*. Sini grabs the rope and clips in.

'Do you really want to go?' Frank shouts. 'You were just in the water – are you okay to climb?'

'It's all right, I'm fine. I feel good.'

Frank nods and a second later she pulls and swings out over the water.

'I'm coming after you,' she shouts up at Kruso. 'I'm just behind you!'

Below her the coastguard boat pushes against *Parker*. A Russian officer grabs Sini's rope and starts yanking it, swinging her from side to side. She unclips the safety knife from her harness, reaches down and cuts the rope beneath her. The officer stares at the rope falling into the bottom of his boat in a little heap. He pulls a gun. Sini looks down, she can see it. The guy's pointing the pistol at her and shouting in Russian. Adrenaline surges through her body, her arms wrench her up the rope, as far away from that gun as she can get.

The coastguard boat is ramming *Hurricane* now. The officers are still eyeing Kieron's camera; it's obvious they want to seize it, they've already grabbed at it four or five times. Frank makes the call to get Kieron and his footage back to the *Sunrise*. 'Kieron, we're coming to get you!' *Parker* swings around so Frank is five metres from *Hurricane*. Frank shouts, 'It's time to go!' Kieron unclips the camera and throws it over the water. Frank fumbles it but manages to keep hold. 'That's great,' he shouts, 'but I need you too!'

By now the gap between the boats is about a metre and the waves are washing them up and down in a deep sweep. Kieron screws up his eyes and hurls himself over the water, falling into *Parker*. Frank slaps him on the back as the driver opens the throttle, the bow lifts and they tear away from the platform and towards the safety of the *Arctic Sunrise*, leaving the other RHIBs to watch over Sini and Kruso.

Suddenly the climbers are being pummelled with water. It freezes their brains and seizes their limbs. The platform workers are using high-powered jets to spray Arctic water over them. The higher they climb, the more pressured the water is and the harder it is to see or feel or hear anything. Sini is just below Kruso on the rope now, but the water is incessant. Freezing. She pulls out a banner – 'SAVE THE ARCTIC' – but it attracts multiple direct jets and disappears in a riot of spray.

They each have a VHF radio plugged into their ears. Anthony, still below them, is looking up, gripping his own radio, convinced they have to get out of there. He shouts, 'Just get back down, get back down quickly!' But the climbers can't hear him, they're being hosed in the head. Even things that are attached to them are flying off in the torrent of water.

Sini can feel Kruso shaking. She's known him for more than a week, long enough to know he's not scared, that this is early hypothermia. Then *bang bang bang*. Gunshots. The guards in the RHIBs are firing over the side into the sea a metre from the Greenpeace boats. The activists are hit by the splash from the bullets. Anthony grabs the radio and cries, 'Shots fired! Abort abort, move away.'

Above them the climbers are trying to descend, but because Sini cut the line when the coastguard was swinging it, the rope now doesn't reach the water. They have to attach a new line to the rope they're hanging off, all the time under the cascade of freezing water from the platform workers above them. Eventually Sini descends far enough for the Russians to forcibly grab her and pull her into their boat, and a minute later Kruso's next to her.

The Greenpeace RHIBs are bobbing in the water a hundred metres away. Suddenly a coastguard officer pulls a gun and fires over their heads. Anthony shouts, 'Go go go!' and the boats swing around as two more shots are fired. 'We need to go, we need to go!' And the activists' RHIBs rip out into the sea.

A few minutes later they're piling into the hold of the *Sunrise*, pulling off their helmets, unzipping their drysuits.

'Fucking hell, did you see those guns? It was crazy out there.'

'What the hell just happened?'

'Did they shoot at you? I thought I saw them shoot.'

'What happened to Kruso? Is Sini okay? We saw her fall in.'

'They came down. They're safe. We stayed out there till they were down.'

Sini and Kruso are taken to the *Ladoga* and marched onto the deck. It's swarming with armed men. Kruso is ordered to kneel, hands behind his back. Sini falls down and hugs his shaking body. She holds him as tightly as she can. A soldier reaches down and pulls at her drysuit; she holds Kruso even tighter but the soldier wrenches her away.

Sini is marched across the deck and pushed into the mess room. She waits to be reunited with Kruso but soon realises they've taken him to another part of the ship. A guard brings her two big blankets and offers her a cup of tea. As she sips from the mug she listens to the ship's internal radio on a speaker and hears the captain of the *Ladoga* issuing commands to his crew. She can't understand what he's saying, but she can tell he's angry.

On the bridge of the *Arctic Sunrise* Dima has the radio receiver at his mouth. 'You have illegally detained two members of our crew. We demand that you return them to us immediately.'

'Heave to and take on board our inspection team.'

'We have absolutely no reason to let you on board. We're in international waters, you have no jurisdiction here.'

'You are in Russia's Exclusive Economic Zone.'

'Well, that's right. So if you suspect us of illegal fishing, please let us know. Because that's the only reason you can legally come on board our ship. Unless you think we're pirates.'

'If you do not submit to inspection, we will use all means at our disposal.'

'You are not allowed on board. We are in international waters.'

'We will use all means at our disposal, including warning shots at your vessel.'

Dima looks at Pete Willcox, the captain of the *Arctic Sunrise*.

'Warning shots,' says Pete, shrugging. 'Okay, let's see.'

The coastguard vessel is coming closer, and through his binoculars Dima can see the Russians taking the cover off a cannon at the bow of the ship.

'You will be shot at unless you immediately stop.'

'Officer,' says Dima, 'I want you to think very carefully about what you have just said to me.'

In the mess room on the *Ladoga*, Sini has been listening to the increasingly demonic shouting on the internal radio. Suddenly there's a bang and the ship shakes. Her tea sloshes in the mug and the surface breaks with ripples. On the *Arctic*

Sunrise the activists see the muzzle flash, there's a burst of smoke and a thud overhead.

'Shit!' cries Dima. 'They're actually shooting!'

THREE

The Russian coastguard keeps up the barrage, firing three shots into the sea beyond the *Sunrise* then demanding the activists take on an inspection party. The firing sounds like distant drum beats and each shot is accompanied by a little puff of smoke from the barrel of the cannon. Three more shots, another warning, then more shots – live shells that explode in the distance. Then around lunchtime, it ends. Silence. All afternoon they wait nervously for the firing to start again. But nothing. As the low Arctic sun dips below the horizon, the crew stand on deck and stare at the *Lagoda*. Somewhere on that ship their friends are being held.

Alex Harris retreats to her cabin, sits at a laptop and writes an email to her family back home in Devon. She's a 27-year-old British climate change activist who's lived in Australia for four years. Her parents knew she was sailing to the Arctic, but no more than that.

> Just wanted to let you know that I'm well and safe.
> I'm not sure if you've seen the news but our activists
> attempted to climb an oil platform from RHIBs. The
> Russian coastguard got pretty violent, and started
> shooting guns in the air and water so we turned

back. They are now holding two of our activists
on their ship. I am perfectly safe, I have been away
from the action, on the ship three miles away from
the platform. We will stay beside the ship until they
release our activists.

When morning breaks the engineers start work on repairing
the battered RHIBs. The others spend the day in the ship's
hold, painting a huge banner demanding the release of Kruso
and Sini. Tomorrow they plan to fly it from the back of a
boat and circle the oil platform and the coastguard vessel.
By the time dinner is served by Ruslan Yakushev, the ship's
Ukrainian cook, the banner is drying and the plans for the
next day have been agreed. The activists file into the mess and
queue in front of a serving counter before taking their food to
one of the long tables. It's just gone six o'clock in the evening.

Frank sits down and glances out of the porthole. The sea is
turning orange as the sun sits low over the water. A full moon
is hanging in a clear blue sky above the *Prirazlomnaya*. He
pokes at his meal with a fork then looks through the porthole
again before deciding dinner can wait. The colours outside
are too beautiful to miss. He drops the fork, pulls on his
sweater and walks out onto the helideck.

Frank breathes in the air, pushes his hands into his pockets
and feels the cold bite of the Arctic on his face. The coastguard ship is three miles away across the blazing water, the
muzzle of its cannon now covered.

Today the sea is flat calm. He wishes it had been like this
yesterday morning, these conditions are ideal for a boarding.
He kicks a chip of paint on the deck and squints his eyes.

Then from behind the *Ladoga* he notices a black dot moving slowly to the left.

It's tiny at first, a little speck that's hard to pick out, but it's getting bigger, changing direction, like a wasp buzzing in front of his face, and from somewhere distant he can hear the low hum of a motor. It's getting bigger, that dot, and staying low to the water, its outline clear against the light blue sky. And it's heading straight for the *Sunrise*. Frank is standing motionless on the deck, his eyes fixed on the dot as he pulls his hands from his pockets and brings them slowly up to his face. Then he cups them around his mouth, turns to the rear window of the bridge and screams a single word.

'Helicopter!'

In the mess room Phil is watching the same speck crossing the porthole glass. He doesn't say anything to Camila and Kieron – who are eating with him – instead he watches it with a curious detachment as it gets bigger and bigger. Then suddenly the sound of conversation and scraping cutlery is interrupted.

'Helicopter!'

And again, this time from a different direction.

'Helicopter!'

The word is echoing around the ship, resonating through walls, shouted in different accents as boots start stamping on stairways and people push their plates away.

'Helicopter!'

Frank is standing on the H of the helideck, watching the chopper swinging around the *Sunrise*, the sound now deafening. His hat flies off his head, his boots slide and he has to lean into the force of the draught to stay on his feet. The side of the chopper is open, a helmet appears and Frank can see a man's face looking down at him. The man drops a long rope that fizzes and zips as it piles up on the deck. A leg swings out of the helicopter, then another. Two big boots hang motionless for a second then an armed commando slides down the rope and lands right in front of Frank. The soldier unclips from the rope, Frank dances in front of him with his arms in the air. More people are with Frank now, maybe five activists, all with their arms raised. Phil is on the helideck, pointing his video camera at the chopper.

Kieron's running down a corridor in his flip-flops and a moment later he's on the deck. And it's just there, a few metres above him. He's stood underneath it. It has a big red star on the bottom, his ears are splitting with the noise and all he can think is, wow, this is amazing, this is the best thing I've ever filmed; I just have to keep hold of the camera long enough to capture it.

On the bridge Pete Willcox is trying to manoeuvre his ship out from underneath the chopper, but the icebreaker is clumsy and slow compared to a helicopter. Throughout the ship the activists are locking doors, screwing portholes closed, blocking every entrance.

Dima is out on deck now, running into the rotor draught, shielding his face with an arm. He can see masks looking down through the open side door, uniforms, big guns, professionals. And they're yelling, gesturing, but he can't understand them. And then *zzzzzzzip!* Another trooper comes down. Dima thinks it's a young kid, maybe nineteen or twenty, but his face is masked. The commando drops to his knees, unclips the rope then raises the barrel of his rifle. He's yelling in Russian, Dima thinks he's saying, 'Get down!

Get down!' but the engine smothers everything, the force of the rotors makes it hard even to stay standing. The kid stabs the air with his rifle; another trooper lands on the deck, and another, and another.

Heavily armed commandos are flooding the ship now. Frank and Dima make a run for the bridge. They know they need to defend it if they're to stop the soldiers taking control of the *Sunrise*. Two of the troopers break away and chase them. Frank reaches the stairs first; the soldiers barge past Dima and throw Frank to the ground outside the bridge door. Dima hears more boots thumping behind him – *boom boom boom* – then he feels a hand on his shoulder pulling him back. He stumbles and falls on top of Frank. A boot kicks him in the side and another boot stamps into his back, squeezing the breath from his lungs. Beneath him Frank is yelling in pain as more boots go in. Dima twists his neck and looks back. On the helicopter deck his friends are lying down, commandos are standing over them pointing their rifles at their backs. And all over the ship, from bow to stern, the *Arctic Sunrise* is swarming with soldiers.

'Frank, are you all right? Are you all right, Frank?'

'Yeah, yeah, yeah.'

'Nothing broken?'

'No, no.'

'I think they're FSB. Special forces.'

Below deck Faiza Oulahsen – a 26-year-old Dutch climate change campaigner – is bashing numbers into a satellite phone. She makes a connection with London, hears the voice of Greenpeace oil campaign chief Ben Ayliffe and shouts, 'We're being boarded!', before slamming down the phone and grabbing the two laptops on the table in front of her. She opens one and presses hard on the 'off' button, but the screen stays lit. She loses patience so starts pressing other buttons, trying to force it to shut down before the Russian security services can gain access to the entire encrypted email history of the campaign and the planning of the protest. She slams the computer closed, scoops up both laptops, runs down to her cabin and slides them under a duvet. As she rushes back into the corridor she bumps into Alex, who's heading for the radio room.

'Alex! We're being boarded!'

'I know!'

'Alex, armed commandos are storming the ship.'

'I know!'

Faiza pushes past, Alex watches her disappear around a corner then she runs down the corridor, throws open a door and is nearly blown off her feet by a Russian military helicopter disgorging soldiers. For a moment Alex stands there with her mouth open, hands over her ears, then she slams the door shut, locks it from the inside and rushes to the radio room. That's where they'll upload footage of the boarding to a server in Amsterdam, but they need to do it before the security forces cut off their communications.

When she gets there, Colin Russell – the 59-year-old Australian radio operator – is waiting. Russian activist Roman Dolgov is sitting next to him, watching the scene on the helideck on a laptop screen through a webcam. Colin slams the door and starts activating a series of lock mechanisms and steel bars designed to give the communications team the few crucial extra minutes of freedom they need to upload the footage. Footage they don't yet have. Alex falls into a chair.

Through a porthole she can see masked men running past with guns, looking in.

Kieron and Phil have retreated from the helideck with their cameras. Phil is trying various doors, looking for a way into the ship. He needs to get to the radio room and hand over the memory card in his camera. Kieron is running for the stairs to the upper deck but he slips on the first step, loses a flip-flop, stumbles and falls. As he's getting to his feet he sees the Russian photojournalist Denis Sinyakov being tackled and thrown to the ground by a commando, his camera sliding along the deck, his arms twisted behind his back.

Faiza is on the bridge now. Through the window she can see Dima on top of Frank. The soldier standing over them raises the butt of his rifle. He's about to smash the window. Inside the bridge Pete glares at him, waves a finger and says, 'No, no, no. I don't want any windows broken, not on my ship.' The commando's rifle hovers over his shoulder. Pete walks towards him, flicks the lock and lets him in. Soldiers stream through the open door.

For a man with the life story of Pete Willcox, this is just another day at the office. He's being boarded at sea by armed men. It's not the first time and it won't be the last.

Frank groans, Dima rolls off him, they help each other up and limp into the bridge. Inside there are five, maybe six troopers, and as many activists. On the deck below, Kieron is standing behind a crane, watching commandos running past, his camera held behind his back. He edges along the side of the ship towards the porthole of the radio room. Phil is already there, clasping his camera, banging on the glass. He nods at Kieron. 'Did you get it?'

'I got it, yeah. Unbelievable footage. You?'

'Gold dust. It was nice of them to turn up at sunset and make it look like *Apocalypse Now*.'

Alex is undoing the screws on the porthole but she can't get the damn thing open. She's unscrewing and unscrewing but the window won't budge. Soon Phil and Kieron are looking at her with eyes wide open, like, hurry up, what the hell are you doing? Roman's helping her and he's a big guy, they're both trying to open the window but it's not happening.

The sound of stomping boots can be heard all around them. Kieron wants to scream, 'Jesus Christ, Alex. Just open the fucking window!' But he can see her face, a picture of pure frustration as she strains at the screws. He takes a deep breath, pushes the camera into his underpants and walks away. Alex motions for Phil to go down the ship to the next porthole, where the campaign office is. Phil checks it's clear and skips down the side. Alex unlocks the radio room door, checks for troopers then runs to the office. Phil is waiting for her on the other side of the porthole. Alex undoes the screws, flips them off and starts opening the window when suddenly out of nowhere a commando appears behind Phil.

Phil sees him and waves his hand, shouting, 'No, no, no!' Alex slams the window in his face and furiously screws it shut. Phil darts away, and because he doesn't have a belt on he can ram the camera straight into his pants. He shoves it right down there and just walks away from the soldier. But the commando pursues him, saying, 'Camera, camera, camera.' Two other soldiers approach from the other direction. Phil stops. He's surrounded.

One of them yanks Phil's coat, spins him around and frisks him. But the soldier doesn't go near his underpants and the large lump of digital equipment nestled between his upper thighs. Phil squeezes his legs together. He knows he won't get to the radio room now, his footage won't be on the TV news tonight, but he's not about to give up his camera card.

On the bridge eight masked soldiers are disabling the communications systems of the *Arctic Sunrise* – VHF radios, GPS tracking, satellite telephones. When the soldiers can't find a button to switch something off, they simply yank on cables and rip them out. A commando is standing guard at each of the outside doors, and there's another one on the inside staircase that leads from the bridge to the ship's lower decks. And all the time an officer is pacing back and forth across the length of the bridge, speaking into his own radio in Russian, taking and giving instructions. Dima speaks Russian, he can understand what they're saying, but not what it means.

'Fifteen to ninety-four, are we doing the seven nine yet?'

'Fifteen, affirmative. Seven nine complete.'

Faiza pulls an iPhone from her pocket, flips on the camera and casually holds it out in front of her. One of the soldiers glances at her, and even through his mask she can see he's smiling at her. It's the one who kicked Frank. Faiza locks eyes with him, she takes a guess at his age – nineteen maybe – and holds his gaze. He's definitely smiling behind the balaclava, those eyes are beaming, he's staring at her. Then he lifts his chin, and in heavily accented English he says, 'Hey, is that an iPhone 4 or an iPhone 5?'

Faiza looks around. The commandos are pulling clusters of wires from the control panels or standing in the doorways with their rifles raised. She looks back at the guy and gives him a look that says, 'Are you serious?' Then she coughs and says, 'Umm, actually this is an iPhone 4.' The trooper nods, and even in the extraordinary circumstances in which they find themselves, even with the mask and the gun, Faiza can read his body language, can feel his condescension. Then through the little hole in the mask she sees his lips purse, and he says, 'Yeah, well, I have an iPhone 5.'

By now a RHIB from the coastguard vessel has delivered a team of senior officers to the *Sunrise*. They're led by a tall man with a thin moustache and three big stars on his shoulder – a captain first rank, Dima thinks. He marches onto the bridge accompanied by a translator and trailing a tail of lesser-uniformed minions, but the commandos barely take note of his presence. He makes a declaration in Russian that appears to impress only himself. The translator says, 'Your ship has been seized. You are accused of attempting to take over the *Prirazlomnaya*.'

The troopers are swarming through the inside of the ship now. In the radio room Alex, Colin and Roman are staring at the inside of a locked door, through which they can hear bangs and thumps that are increasing in volume, and the shouted demands of Russian commandos. 'Open this now! Open this now!'

Alex doesn't open the door, instead she opens a laptop and tweets from the account @gp_sunrise to thousands of people around the world.

Russian authorities onboard with guns. They are breaking into the comms room now. #savethearctic

Then again.

> This is pretty terrifying. Loud banging. Screaming
> in Russian. They're still trying to kick in the
> door #savethearctic

'Open this door! Open now!'

Alex looks at Colin, Colin looks at Roman, Roman looks at Alex.

'Do we open it?'

'I don't know. I mean, they sound a bit demented.'

'I don't think they need us to open it. They'll be through it soon anyway.'

Alex finds a piece of paper and scribbles 'SAVE THE ARCTIC' in big letters then holds it up, facing the door – a personal message to the commandos who are about to break through. It only takes a moment more. Colin stands up and spreads his arms to shield the other two. *Bang bang*, and then *bang*, three whacks and the whole door comes off its hinges and commandos surge into the room.

Alex, Colin and Roman are marched down to the mess room. When they get there Alex sees her friends corralled and heavily guarded. She looks at their faces and sees a mixture of fear, anger, boredom and disorientation.

On the bridge the Russian officer is engaged in a protracted argument with Pete. The man is demanding that he sail the *Arctic Sunrise* to Murmansk on the Russian mainland, while Pete is politely refusing to co-operate in any way. The officer huffs, he expresses profound displeasure in Russian and broken English. 'This you say no, yes? But this … you must say yes.' Then he informs Pete that he will now order the *Sunrise* to be towed to Russia by the *Ladoga*. Pete shrugs. For him it's a matter of deliberate non-co-operation. They've boarded his ship illegally, they have no right to be here and he isn't about to make things easier for them.

The crew on the bridge are herded down the stairs and into the mess, all except for Pete, who the officer hasn't finished with yet. The door to the mess is guarded by two armed commandos. On a blackboard facing the entrance, the activists have written: 'Russian soldiers, welcome to the Arctic Sunrise!'

FOUR

Across the ship telephones have been seized, radios disabled and Internet access shut down. But in the mess room, where the crew are being held – and unknown to the occupying Russians – a single telephone is still working. It's a black plastic handset connected to the last functioning satellite link. The first activists to be pushed into the mess managed to hide the phone from the troopers and now it's in the galley – the ship's kitchen – where the smokers have converged to exhale up a ventilator.

Frank calls the Greenpeace office in London. He's whispering in precise little sentences. He says there were twenty commandos, heavily armed, guns and knives, all wearing masks. The ship is being towed to Murmansk, he says, and it'll take four or five days. Some of the troopers are talking about serious charges, time in jail, but he thinks that's bullshit. Then he says he has to go, he doesn't know how long they'll have the phone for, and with that the line goes dead.

The trooper guarding the door shifts the weight of the rifle cradled across his chest, looks away then looks back again. Behind him more heavily armed men are stomping through the corridors, going from cabin to cabin, searching bags, drawers, tins, everything. They're coming out carrying the activists' books, computers, soap bags. Frank sees one of them clutching his bottle of Sailor Jerry rum.

Minutes pass, then hours. Groups form around card games. The smokers execute a complete takeover of the galley. Phil pulls the camera from his underpants and stashes it in the extractor fan. It's a relief. There was a stiff plastic cable tie on the camera that was cut off diagonally and it was digging into his thigh.

Over on the *Ladoga,* Sini is sat on the edge of a bed in a locked cabin. She's not seen Kruso since they were arrested. She's been here nearly two days. The Russians have been pleasant enough, but she wants to be back with her friends.

Sini Saarela has been an activist since she was a teenager. For years she's been scaling highly polluting fossil fuel infra-structure. Her mother would ask her, 'Why does it always have to be you?' and Sini would reply, 'Who else is it going to be?'

She feels a strong connection to the Arctic, she grew up here, she spent time living with the Sami people in the far north and conducted forest mapping in Finland, Sweden and Norway. When someone asked her why she was joining the *Arctic Sunrise* she said in a soft melodic Finnish accent, 'Because it's our Arctic. Who are these Dutch and Russian companies, coming up here and messing up our Arctic? It's humanity's Arctic. It's my nature, the nature where I grew up.'

There's a knock on the door. It opens and an officer is standing above her. 'Okay,' he says, 'you're going back to your ship.'

She's reunited with Kruso, she throws her arms around him but a moment later they're pulled apart. A soldier gives them back their drysuits and life vests. They're put into a RHIB and driven across the water to the *Arctic Sunrise*. As they come closer Sini sees there's no one on deck to greet them. The Russians drive the boat alongside the Greenpeace ship. They order Sini and Kruso to climb the pilot ladder. And Sini's thinking, I know I didn't climb well yesterday, I wasn't happy with it either. Okay, so the protest only lasted a few minutes, but was it so bad that there's no one here to meet us?

But when she pulls herself up the last rung of the ladder she sees soldiers with heavy guns and balaclavas. And then she understands.

They're taken down corridors, past masked men coming out of their friends' cabins carrying bags and computers, and then they're pushed into the mess room. There's a moment of silence as they walk in, then the crew surges towards them, hugging them, some crying with relief. They're together now. Together on the *Arctic Sunrise*. All thirty of them.

One of the troopers stands in the doorway and asks for silence. He announces that the activists are to be taken one by one to the laundry room to be searched, and asks for their co-operation. Quickly phones are slipped out of pockets and hidden under cushions. Phil eyes the extractor fan. He's worried it will be searched and he'll lose the footage. He doesn't need to hide the whole camera, just the thumbnail memory card.

Surreptitiously he pulls off a boot. He pulls out the foam sole and with a kitchen knife he cuts a little slot in the heel. He strolls into the galley, looks over his shoulder then pulls out the camera. A commando is standing just two metres away from him, looking in the other direction. Phil's heart is thumping in his ears. He takes a step to the side so his back is facing the trooper, then he slips the card into the sole, shoves the camera back in the extractor fan then bends down and pulls his boot back on.

A trooper enters the mess and folds his arms across his chest. 'Okay, listen up! This is the deal ... ' Phil spins around. The man is speaking in Russian but he's got a translator standing next to him. 'You will be allowed in here and the lounge. You will only be allowed in the corridor of this deck and the deck above. You're not allowed to go outside and you're not allowed to go into the hold. I know some of you have cabins there. I don't care, you can't use them. For now you stay in here. You don't leave this room.'

'Can we go out on deck to smoke?'

'No, you're not allowed to go out on deck. You're only allowed in these two corridors.'

'Can we see Pete, the captain?'

'No, you cannot see the captain.'

'Are we going to Murmansk?'

'Yes.'

'What will happen when we get there?'

'I cannot say.'

'Will we be able to go home?'

'I cannot say.'

'Is there anything you *can* tell us?'

'I cannot say.'

In Copenhagen, Moscow, London and Amsterdam, Greenpeace staff are calling relatives of the crew. Some families express shock and horror, others sound like they were expecting this. In Amsterdam, Mimount Oulahsen-Farhat – Faiza's mother – hears the telephone ringing. It's a Thursday, she's at home watching television. The man on the line says he's from Greenpeace. He says the ship her daughter is on has been stormed by Russian soldiers. Everything is okay, he says. Everyone's safe. When there's more information he'll contact her immediately. Mimount puts the phone down, stares at the wall and bursts into tears.

At just past 10 p.m. Moscow time, Russian state media reports that the *Arctic Sunrise* is being towed to Murmansk. Five thousand miles away in Washington DC the first protest is beginning, outside the Russian embassy. By the next morning there will be demonstrations planned or happening in thirty cities across the world.

On the *Sunrise* the night stretches out before them. Under the watch of armed guards the activists spread out across the mess, playing card games or lying on the floor under the tables. Only in the morning – more than twelve hours after the raid – are they finally allowed to leave. An officer stands in the doorway and reads out a list of numbers. 'These are the cabins you are not permitted to enter. The others you can use. You can sleep now, if you want.'

They spill out into the corridor and retreat to the cabins. Frank tapes his computer memory stick to the bottom of the table in his cabin. Kieron hides his camera card in the light fitting in the ceiling. Fazia slips a mobile phone into the underside of a drawer. Alex slides hers into a bag of rice.

After a few snatched hours of sleep they start reappearing from their cabins and gather again in the mess. Dima passes three of the commandos in the corridor. They're still masked but he can see their eyes are red, and they're using the tips of their gloved hands to steady themselves against the wall. He can smell burnt alcohol on their breath.

The *Sunrise* hits some weather, she's not the most stable ship and the troopers stumble down the corridors, seasick and hungover, guns slung over their shoulders. When the activists try to talk to them the soldiers are impatient and angry.

Dima tells the other activists he's not worried about their situation, he says the rest of them should be as relaxed about this as he is. This is the third time he's been arrested in these waters. The first time was in 1990 when it was still the Soviet Union. He says the ship will be taken to Murmansk, they'll be held for a day or two, there'll be some paperwork to complete then they'll sail out of port and back to Norway, where they left from last week. And in the meantime Greenpeace offices around the world will be exploiting the media moment to the maximum. He's been getting arrested on the high seas for a quarter century. This is nothing to get stressed about, says Dima. Best to just sit back and chalk it up as another great story.

The main threat now is boredom. The activists play endless games – chess, Scrabble, Pictionary, Mexican poker. Anything to eat up time. Pete Willcox, the captain, is allowed down twice a day for lunch and dinner accompanied by a guard before being taken back to his cabin.

Three days in and Frank is thinking these Russians are real professionals. Not arseholes. Just doing their job. A commando even helps Frank carry a tray with tea and cakes up the stairs, the commando's face covered by a balaclava, a Vintorez special forces rifle slung over his shoulder and a handgun and a knife strapped to his thigh.

When the crew goes to bed a contingent of troopers prowls the corridors or stands guard outside the cabins. Camila Speziale wakes in the night needing the bathroom. She pushes open her door and finds herself face to face with a masked man cradling a rifle. 'Hi,' she says then eases past him and slips into the washroom. Two minutes later she walks past him again.

'Hi again.'

'Hullo.'

'Goodnight then.'

'Yes.'

The next day Camila tries to engage the soldiers in conversation, but her advances are met with surly monosyllabic replies. She resolves to orchestrate a detente. It's four days after the boarding and she decides to make them pancakes. She prepares them with a chocolate and fruit topping. Dima tells her the Russian word for pancake is *blini*, then she approaches the guard at the mess room door.

'*Blini*. Tasty. You want?'

'*Nyet*.' The trooper stands ramrod straight, his eyes staring over Camila's shoulder.

'Chocolate and fruit.'

'*Nyet*.'

'For you, *blini*.'

He looks down at her, smells the pancake, then looks away. '*Nyet!*'

FIVE

At the Greenpeace office in Moscow, a ping pong table has been rolled into the main meeting room to accommodate the numbers of people working on the campaign to free the *Sunrise* crew. The room was requisitioned and turned into a crisis response centre the morning after the ship was seized. A light-fixture resembling a chandelier hangs from the ceiling and the floor is a smooth polished wood, so the team dubs it the Dance Hall. But right now it feels more like a bunker. Twenty staff crowd in from six in the morning until past midnight, organising lawyers for the moment when the ship arrives in Murmansk.

But from the first moment they're swamped by a full-frontal propaganda assault. The protest was a terrorist attack, the activists are CIA operatives, they were acting as stooges for Western oil companies, the pod could have been a bomb. The lies come from all corners of the Russian establishment – from journalists, ministers, the security services, and from the state-owned oil company, Gazprom.[5]

In Amsterdam – where Greenpeace International is based – the organisation's digital campaign team is looking to mobilise global public opinion. A conversation on Skype sees the first use of a phrase that will soon become the name of an international drive for the crew's freedom.

James Sadri: we want to go for a big push on
#freethesunrise30 as a hashtag to mobilise people

Andrew Davies: #savethearctic

Andrew Davies: It keeps arctic in the frame

James Sadri: #freethearctic30

Andrew Davies: #FreeTheArctic30

James Sadri: nice

Meanwhile, Greenpeace legal chief Jasper Teulings is working with Moscow to assemble a team of lawyers for Murmansk. From the first moment it's clear to him that the organisation is in serious trouble. He's a lawyer himself but he knows this isn't about the law, and that's what scares him. Greenpeace is facing what he calls 'a lawless, cowboy situation'.

He telephones his colleague Daniel Simons, a Russian-speaking lawyer who's on a romantic holiday in Venice, and asks him to fly to the Russian Arctic immediately. Next he contacts the foreign ministry in Amsterdam (Greenpeace ships sail under the Dutch flag) and pushes them to bring a case before an international maritime court to demand the release of the *Sunrise* and her crew. But this is the official Russian–Dutch year of friendship: huge trade deals are planned, a state visit to Moscow by the Dutch king is just weeks away.[6] Surely Teulings can't expect all that to be put at risk over an Arctic oil protest?

Two days after commandos raided the ship, a man claiming to be a reporter turns up at the Moscow office asking for a tour of the building. Staff there soon become suspicious. It's the way he's dressed and the questions he asks when he interviews them. 'And who ordered the protest at the platform?' 'How does your hierarchy work?' They think the man is probably from the Federal Security Bureau. FSB agents have a certain style and this guy is an archetype. He's wearing a white shirt and a black leather jacket, and he has what one campaigner calls 'a Bill Gates type of haircut'.

The man says he's going to the toilet then disappears. A few minutes later he's found wandering alone along a corridor. Eventually he leaves. And from that moment onwards the Moscow team discusses sensitive issues on an outdoor balcony. They're convinced they've been bugged.

In Copenhagen the executive director of Greenpeace's Scandinavian operation, Mads Christensen, has been handed leadership of the global campaign to free the arrested activists. Christensen is forty-one years old, the son of a cinema owner, a graduate in political science and the only national leader in the Greenpeace world to have started in the actions department – the team that organises and executes protests. Two years ago he jumped into the water in front of an icebreaker to delay its journey to join Shell's exploratory Arctic drilling operation off Alaska. He is tall, slim, with blond hair and black, thick-rimmed glasses.

Straight away it's clear to him that Greenpeace is in 'deep, deep shit'. This is going to need an international campaign of indefinite length. He spends a day putting a team together, recruiting experienced staff from across the organisation, and the next morning they walk away from their old jobs and devote themselves to the release of the Arctic 30.

Christensen's first, most urgent task is to get organised in Murmansk before the *Sunrise* docks. The crew is going to need more than just lawyers. Appointed to lead the ground team is 38-year-old Belgian Fabien Rondal, a Russian-speaking former roadie for Rage Against the Machine.

'You need to get there as soon as possible,' Christensen tells him. 'We're thinking the ship will be there Tuesday or Wednesday and we need you and your team in place before then. I can't tell you when you'll be able to come home. Nobody knows what happens next.'

SIX

Alex Harris pulls a mobile telephone from the bag of rice in her cabin where she hid it, and dusts it down. She looks over her shoulder then turns it on and watches the screen. Nothing. It's taking for ever. She shakes the phone. Still nothing. Then the screen lights up and her heart leaps as she sees one bar of signal.

It's five days since commandos stormed the *Arctic Sunrise* and now, finally, there is a dark shadow on the horizon. Land. The working phone in the galley has long since died, but the mobiles they hid in the cabin are getting reception. Alex dials a familiar number and hears it ringing. Her parents can barely believe they're speaking to her. Breathlessly they tell her there are camera crews on the doorstep and pictures of her in the newspaper.

The crew take turns to call their parents, their partners and kids. They tell them they're okay, that this will soon be over. Frank telephones the London office. He says the soldiers were flown in specially from Moscow, these guys aren't amateurs, they were operating under specific orders. 'I've been on a lot of actions,' he says, 'but this feels different. This isn't good. The Kremlin's up to its neck in this.' But then he says, 'Something's happening, I've got to go, I'll try to call again.' And the line cuts out.

They've come to a halt. The *Arctic Sunrise* is anchored and the coastguard vessel ties up alongside. A Russian officer – a new guy – walks into the mess room with a translator. 'Okay, listen up! You people are going to be taken off the ship in two groups and interrogated, so prepare to leave. Two groups, fifteen people then fifteen people. Group one, you have five minutes to get ready.'

'What should people bring with them?' asks Dima. 'People don't have their documents. You've confiscated their papers.'

'It's cold, so you should wear warm clothes. But you'll only be gone for a few hours, so don't bring too much. And don't worry about your passports, you won't need them anyway.'

Sini asks, 'I have to take a medicine. How much of it do I need to bring?'

'You should bring enough for one full day. Twenty-four hours. You're not going to need it for that long, but just to be sure.' The interpreter translates his words, then the officer adds, 'Actually, just to be safe, bring enough medicine for three days. I'm sure you won't need it, but still, bring it for three days.'

Dima has well-attuned bullshit antennae and right now they're twitching. He's been around long enough to know when he's being lied to. He goes to his cabin and packs everything. He packs all his clothes, pants, all his T-shirts, underwear, socks, a thick book and extra chewing tobacco. Everything. All of it in the big pink bag he brought from Sweden when he boarded the *Sunrise*. And now he's thinking, shit, I'm going into one of the most homophobic countries in the world with a huge pink bag, from a ship with a rainbow on the side. Great.

The first group of fifteen is taken up onto the deck. It's the first time they've been outside for days. The ship is swarming with armed uniformed men, they're covering every inch of railing, dozens of eyes stare at the activists as they shuffle along the deck. The first fifteen are transferred to another, smaller ship that's also heaving with soldiers. The journey takes forty-five minutes, and when they pull into Murmansk dozens more armed men are waiting for them on the jetty.

It's dark now. Raining. Faiza Oulahsen – the young climate campaigner from Holland – looks around. The port is decrepit, with rusting cranes towering over an old bus parked up with its door open. A hand touches her shoulder, she turns around, a woman in uniform is facing her. The woman says, 'You can wait on the bus if you'd like to.' Faiza shakes her head. She hasn't felt the wind on her face for five days and she's savouring the rain and the cold, stiff breeze.

It's the last time she'll feel it for a very long time.

When all thirty have been brought to land they're ordered to board the bus. They gingerly take seats then watch it fill up with soldiers who sit next to them, behind them, in front of them. The soldiers' faces are covered by black masks with little holes for their eyes and mouths. The bus is old and smells of metal. They can taste it.

By now the video journalist Kieron Bryan is worried. He has a reporter's eye for detail and since they docked he's counted 300 uniformed men with guns. Some from the FSB, some from the police, some from the army. There were at least sixty on each ship – the *Sunrise* and the *Ladoga* – and another sixty on the ship when they were transferred to land.

Maybe 150 meeting them at the port. Kieron stares through the window at the old Soviet-style buildings and feels a potent charge of fear building in his legs. After a few minutes the bus comes to a stop outside a huge, well-lit building surrounded by scores more camouflaged rifle-wielding men.

Camera flashes burst through the windows. On the pavement, among the military contingent, the crew sees people wearing Greenpeace T-shirts, arms aloft, fists clenched. It's Fabien Rondal and his team. The Belgian was tipped off by a local journalist that the crew would be brought here. There are claps and shouts of support, the prisoners make V for Victory signs through the window, then they're pulled out of the bus and pushed through the gauntlet of lights, cameras and guns, and into the building.

The thirty are quickly processed through a metal detector (keys and cash were liberated long ago, phones were left on the *Sunrise*). Phil nervously waits for the camera card in the sole of his boot to beep, but he passes through the machine without setting it off. They're led up a flight of stairs. Two men in shiny suits are standing at the top. As the activists pass them the men grab their arms and shake their shoulders and in Russian they say, 'I'm your lawyer! Don't sign anything! Demand to see your lawyer! Don't sign anything! Demand to see your lawyer!'

Dima translates the message and it gets passed back down the line. 'Don't sign anything! Demand to see your lawyer!' Then a door swings open and they're pushed into a large room with thirty empty chairs. They are at the headquarters of the Investigative Committee – Putin's domestic legal hammer, the organisation that has put Pussy Riot and scores of other political activists behind bars.[7]

The walls are covered with instructional posters illustrated by scenes of violence and bold banner messages in Russian.

HOW TO TELL IF A PERSON
HAS BEEN KILLED BY A STAB WOUND

And above these posters is a huge framed portrait of Vladimir Putin.

The crew take a seat or kick their heels. The mood is heavy. This feels bad. A uniformed woman appears in the doorway.

'Allakhverdov? Andrey Allakhverdov? Roman Dolgov? Dima Litvinov?'

'Yes.'

'Follow me, please.'

Then a man appears behind the woman.

'Faiza Oulahsen? Sini Saarela?'

Sini is led down a corridor to a small, grimy interview room. It's cold. A woman is sitting at a desk in front of a computer monitor. Sini sits down. Opposite her is a policeman. He starts to speak. The woman translates.

'We have to write a written report. About the incident.'

'The incident?'

'The incident. A serious crime has been committed and we suspect you may be responsible.'

In rooms along the corridor the activists are facing investigators and translators. Dima is sitting opposite a man in scratchy blue trousers and an open-necked civilian shirt who introduces himself as a colonel in the Investigative Committee. He has in front of him a sheet of paper, which he lifts with some solemnity before reading out an official proclamation. 'You are now considered to be a suspect in a case of piracy.'

'Piracy?' Dima shakes his head. For a moment he wonders if he has misunderstood. 'Did you say piracy?'

'Yes, piracy.'

'Are you kidding me? How can you … do I look like a pirate? Come on, man. This is insane.'

The colonel raises a dismissive hand. 'Spare me. You don't need to say anything right now, unless you are proposing to give me a statement?'

Dima screws up his face and pats the air dismissively. 'No. No, I'm not giving you a statement. I'm giving you a fact. You can't charge us with piracy. That's just so dumb it hurts.'

'And that is your statement?'

'No! No, it's not a statement, it's … it's just … seriously, man. *Piracy?*'

'Either you're making a statement or you're not. Make up your mind. In the meantime I need to tell you you're remanded in custody until the court hearing.' The colonel scribbles something on the sheet of paper, and still looking down he adds, 'We're going to ask the court to hold you in custody for the length of the investigation.'

'Well, how long is that going to take?'

The man looks up, lays down his pen and presses the tips of his fingers together.

'We can keep you for a year and a half.'

'What?'

He shrugs his shoulders.

'Okay,' says Dima, 'what's the sentence for piracy?'

'Ten to fifteen.'

'*Years?*'

'Years.'

Dima crumples in his seat. His heels slide on the floor. He grips the handles of the chair and holds on tight, steadying himself. In the very pit of his stomach he can feel a tight knot of fear, like a fist in his belly. And this is the moment Dima Litvinov crosses into a new plane of existence. He was just in the middle of a normal, good, hardcore Greenpeace action, and now suddenly he's facing fifteen years in jail.

Meanwhile, in a room down the corridor Roman Dolgov is listening to an argument between two investigators.

'This is crazy,' says one. 'There are thirty of them.'

'What can we do?' says the other. 'It's coming from the top. An order from the minister.'

They're in the hands of the Kremlin now.

When they've all been read the declaration, they're taken out to transport vans. An officer barks orders in Russian. Dima translates.

'Okay, listen up, detainees! You are being taken to lock-up facilities while you await hearings to consider your continued detention during the course of the investigation. Please, take your places.'

The back doors of the transport vans are thrown open. The activists are loaded into the vehicles, engines turn over and growl into life. Half an hour later one of the vans pulls into a submarine base. Kieron, Frank and Cristian are hauled out and led through the complex to a cell. It measures two metres by a metre and a half. Kieron grimaces then slowly lowers himself to the floor. He's tired, confused, unsure how he ended up here.

The three of them are sat on the floor in a row with their feet up on the wall opposite. In the corner of the cell a human shit is radiating a grotesque smell. It must have been there for weeks, maybe longer. It's green and white. Covered in fungus.

Across town Dima and Pete Willcox are driven into the yard of a police station. They're strip-searched and their finger-prints are taken, then they're put in a cell with a Russian guy, a drunk who's in for assault. The toilet is a hole in the floor behind a partition that you can crouch behind. The Russian is leaning against the wall, smoking a cigarette. Pete lies down and closes his eyes. It's been a long, long week. Dima sits on the floor and rests his chin on his knees. In the background, fuzzy and barely tuned, a radio is playing a Russian pop song, but a few minutes later the news comes on, and the *Sunrise* is the lead story.

' … when there was an attack on the platform by the group, who were pretending to be ecologists … have now arrived in Murmansk … tried to take over an oil platform … extremely violent … injured coastguard officers … ' Dima's head jerks up. In his stomach the knot tightens. ' … managed to appre-hend them … clear case of piracy … the state has succeeded in defending the Russian Federation from this attack. The government is not yet sure if this was an operation directed by the secret services of a foreign country or whether it was a rival corporation paying the group to launch their assault. But obviously this was an attack on Russia's legitimate inter-ests.' Then an official from the Kremlin is interviewed – *the Kremlin!* – who says, 'Let me be clear, this is not something we are going to tolerate. There will be serious consequences for these individuals.'

Dima hugs his knees tight. The drunk draws on his cigarette. 'Ten to fifteen,' he says, blowing out smoke. His words are slurred, his nose is dark red, the colour of cherries. 'That's heavy.'

'I'm sorry?'

'Piracy. Ten to fifteen years, that's heavy, man.'

'You know we're Greenpeace?'

'Of course.'

'You really think we're going to do time?'

'Well, once you're in here, and if they say you did it … '

'But do you really think we'll go to *jail*?'

'If they say you did it, you usually did.'

Dima stares at him. Is this guy for real?

What does he know? He's just a drunk.

SEVEN

It's a week since the *Sunrise* was stormed, two days since they landed in Murmansk.

Across the city the activists are woken by policemen and taken out to prison transport vehicles. They're called *avtozaks* – vans with tiny compartment cells so small that the prisoners' knees are pressed against the wall in front of their noses. And this is how the thirty are taken to court to learn their fate.

The Russian men arrive first – Roman Dolgov, the photo-journalist Denis Sinyakov and Andrey Allakhverdov, the ship's fifty-year-old chief press officer. They're locked in a holding cell. Their lawyer comes in. He tells them Putin has been talking about the case. The President said they're 'obviously not pirates',[8] but Putin also claimed the commandos couldn't have known the *Sunrise* crew were genuine environmental activists, that the authorities had grounds to suspect the campaigners were using Greenpeace as a cover for more sinister motives.

The door opens, a guard appears.

'Sinyakov?'

'Yes.'

'It's time.'

Denis is led away. He isn't worried, he assumes the hearing is a formality and he'll soon be free. At the police station where they were held, the officers went to the shop and bought him toothpaste, a hairbrush and shampoo. And when he left the station for the court this morning they gave him a meal in a box. Cops in Russia are never this pleasant. Denis is confident it will soon be over.

He's locked in a cage in the courtroom. He's familiar with the set-up. As one of Russia's leading photojournalists he's covered the most celebrated trials in his country's recent history. He was only on the ship to photograph the protest, but now it's him in the dock.

The hearing starts. He looks at the judge, a middle-aged guy, fifty-five maybe. And he's seen this judge so many times. Not this man exactly, but this type. The man's eyes betray complete disinterest in the case. It was the same at the Pussy Riot trial. It was just like this in the courtroom with Mikhail Khodorkovsky – the oligarch prosecuted on trumped-up charges after he challenged Putin.

Now Denis is worried.

The prosecutor speaks. Denis's lawyer responds. The judge looks bored. When he does speak – to deliver his verdict – the judge does so without emotion, mumbling his way through a text on a sheet of paper in front of him. There is no trace of feeling in his voice, it's like he's reading from the telephone directory, but his final words explode in Denis's face.

' … therefore the accused is jailed for two months while the authorities investigate the criminal attack on the oil platform.'

Denis grips the bars of the cage. He's staring at the judge, shaking his head. A guard opens the door, handcuffs him and leads him out of the courtroom and back to the holding cell. Andrey and Roman look up.

'Well?'

'Two months.'

Roman jumps to his feet, but before he can process the news, a guard is leading him out of the cell. He's taken to the courtroom, where he's told he will be jailed for two months while the investigation continues, and for ten to fifteen years if he's convicted.

Downstairs, a van holding the women parks at the back entrance of the courthouse, and one by one, their hands cuffed behind their backs, they jump out and are led up the stairs towards the holding cells.

Half an hour later Kieron, Frank and Cristian step out of an *avtozak*, surrounded by guards. One of them flashes a lopsided smile and says, 'Welcome to Russia.' Frank looks around. The place is swarming with heavily armed soldiers and policemen. They really are taking this seriously, he thinks. They're taken to a holding cell. Inside are French Canadian activist Alexandre 'Po' Paul and electrician David Haussmann, whose partner back home in New Zealand is pregnant.

'You guys all right?' Frank asks. 'No offence, Po, but you look terrible.'

Po-Paul shakes his head slowly.

'I've got some really bad news.'

'What? What's happened?'

'It's jail.'

'Oh, piss off,' Frank says, laughing. 'Not the time for jokes.'

'Seriously. We're all getting sent down.'

Frank stares at him, biting his lip, then he looks at Kieron, whose mouth has dropped open.

'I've just been in court,' says Po. 'They've given me two months' detention, minimum. For piracy.'

Frank is ashen-faced. 'Jesus, seriously?'

'Yeah.'

Kieron slides down the wall. He grips his shins and buries his face into his knees, then looks up at Frank and Po-Paul and says, 'Guys, we have to make sure we're in cells together. We can cope with anything then. We can protect each other.'

And Po-Paul says, 'I've read books on this. You've got to go into your cell and if there's a weaker person than you, you have to just beat the crap out of him.'

Kieron shakes his head. Po-Paul's got a dark sense of humour, and right now it's not helping.

One by one the activists are taken from holding cells and led to the courtroom to be told they're going to prison. Alex gulps back tears and rubs her eyes. She's staring at the judge, barely able to take it in. The man has a fat neck, he's wearing tinted glasses and is draped in a black gown with silver buttons below the chin. He's sitting at a raised dais, leaning back in the middle of five high-backed leather chairs.[9] The judge doesn't once look at Alex, instead he stares down at his desk or occasionally looks at the officer from the Investigative Committee, who tells the court Alex could interfere with evidence unless she's jailed. A translator is standing next to the cage, whispering the words to her in English, but Alex is barely listening. All she can think of is her family back home in Devon. She knows how hard this will hit her mother, and she just wishes she could speak to her and explain why she was on that ship.

Next, Camila is brought to court and jailed, then Phil and Kieron. By now the hearings are overrunning. Under

Russian law if a defendant isn't processed within two days of their arrest, they have to be released. The crew were officially arrested at the port on Tuesday evening, and now the authorities are in danger of missing the forty-eight-hour deadline.

Frank is taken from the holding cell and led down a corridor, and as he passes one of the cells he hears a voice shout out, 'You've got some fucking questions to answer, Frank. It's your fault we're getting sent down.' He twists his head but the guard prods him and he keeps on walking, and a minute later he's being locked in a cage in the courtroom. An officer from the Investigative Committee, dressed in a brilliant blue uniform with gold stars on his shoulders, stands up and reads out the case against the defendant. He says there was a violent attack on the oil platform. The authorities had no way of knowing if the activists were terrorists. Then the prosecutor stands up and nods gravely.

'Well,' he says, 'from the evidence, we can see that this is a very serious accusation. Very serious indeed. And it's clear – I have to say, it's very clear indeed – that the evidence is overwhelming.'

Then Frank's lawyer stands up, tapping his watch and speaking quickly in Russian. Frank doesn't understand what's going on, but then the lawyer comes back to the cage and explains. Time has run out, he says. The deadline has passed. It's too late for the judge to send Frank to jail now, but rather than releasing him the judge has beaten the clock by ruling that the hearing should be postponed. Frank will be returned to the naval base then brought back to court in three days.

After him come Dima, Sini and five others. They all have their cases postponed. They'll be taken back to their cells at the police stations.

'What about the others?' Sini asks the lawyer.

'They are going to jail tonight.'

Alex Harris is handcuffed in her holding cell then led down a staircase and outside to a waiting *avtozak*. She's pushed into a little wardrobe cell. She can barely move. She shouts out to see if there's anyone else around. Some of the guys shout back, they're in the van with her. It pulls away and as it turns a corner her face is pressed against the cold metal wall.

They drive for twenty minutes, maybe half an hour, before the van comes to an abrupt stop. The engine falls silent. She can hear dogs barking, doors slamming, voices shouting in Russian. She stares at her shoes and draws a deep breath. A powerful current of fear is running through her body, making her heels bounce and her lungs tighten. 'It's okay,' she whispers to herself. 'It's okay. It's going to be okay.'

And then the door opens and she's led out of the van, followed by the others. They look around and see a dark sky, spotlights sweeping over the ground, high walls surrounding them on three sides, rolls of razor wire.

They're in the reception area of a converted mental asylum. It's the region's main isolation prison. Murmansk SIZO-1.

Alex is led through a courtyard. In the darkness she can make out the shapes of people at the windows. Russian prisoners. She doesn't know if they're crazy but it sounds like they're crazy because they're screaming into the night. She turns to Camila. 'Well, this is home,' she says and laughs nervously, but sheer panic is coursing through her body.

Just behind them are Kieron and Phil. The activists are in a three-sided courtyard with four floors, fifteen windows across each wall. The bars on the windows aren't straight and the windows aren't quite square, and there are ropes hanging down the brickwork, connecting them horizontally and vertically. The ground is littered with sheets of beaten corrugated tin that have fallen from the roof. There are tall weeds growing through gaps in the pavement. The guards prod them forward. A spotlight is following them, casting shadows on the ground in front.

Camila looks back and sees the expression on Kieron's face. 'Hey,' she says, 'it's going to be okay. We'll survive this.' Kieron tries to smile back at her, but he's not thinking it's going to be okay. Kieron's thinking this is a nightmare, this is hell on earth, and he doesn't know how he's going to survive this place. The air is filled with shouting and screaming, but he doesn't know what the prisoners are saying. The cries are guttural, something close to animalistic, and they're coming from the windows, from the throats of women and men. The activists hear the word 'Greenpeace' shouted a few times, which means the Russians know they're coming. And Kieron's asking himself what they're saying before they scream that word. Like maybe it's, 'Let's fucking kill the people from … '

They're marched past a basketball court, down the inside of a fence topped with razor wire and through a doorway. They're inside now. A guard grips one of Alex's wrists and flicks a handcuff over it then spins her around, yanks an arm behind her back and cuffs the other wrist. She's marched down a corridor as far as the last door, where the handcuffs are removed and she's pushed inside a tiny room. A woman is standing opposite her. She's huge, there's only just space in here for the two of them. The ceiling is pink and the walls are covered with mirrors.

'Knickers down.'

Alex stares back at her. The woman is wearing a crisp blue uniform, hair pulled back into a tight bun below a peaked cap, cheeks pooled into a huge neck, no jaw to speak of, a mouth so small and tight it looks like the valve on a beachball.

'Knickers down!'

'Seriously?'

'Knickers.'

'What?'

'Down!'

Alex closes her eyes, takes a deep breath and pulls down her jeans then her knickers. Slowly, tentatively, Alex squats, steadying herself by holding her hands against the walls. Her palms squeak as they slip on the mirrored glass. Awkwardly, with laboured breaths, the guard lowers her broad form until she's on her knees. She's right next to Alex now, bringing her face down, lower and lower, her breaths getting louder until she can look up and inside her.

'Okay, knickers on.'

Alex is taken to the room next door and handed a mug, a bowl, a spoon, a tiny rolled-up mattress and a sheet. She's led back down the same corridor, around a corner and up a staircase. The guard stops her outside a solid metal door, slides a key into the lock, turns it and pushes. And Alex is presented with a room that's two metres by five metres with steel bunk beds that are fixed to the wall. It's green, very green, nothing else in there except a sink and a toilet. She steps forward. The door slams behind her.

Two minutes later Camila is stood outside a cell door on the same corridor, shivering, clutching her bedding. The guard opens the door, and Camila is expecting Alex to be in there. But the cell is empty. All she has is her blanket, her cup, her spoon and her jacket. The door closes. She's alone. She makes the bed, sits down on it and stares at the wall. She's not scared. She's furious. The investigators, the guards, the judge. It's all bullshit. She scrunches up the blanket in her fists and kicks the floor.

On the corridor above her, Phil Ball is being strip-searched. He pulls out the insoles of his boots to show the guard there's nothing in there. That way the guy won't look more closely, he thinks. He has to surrender his shoelaces then he's led down the corridor with Po-Paul and Kieron. The guard stops him outside one door, Kieron outside the next one along, then Po-Paul outside the door after that. They're looking at each other with an expression that says, oh shit, so we're not being held together. Then the guards pull open the doors and a cloud of cigarette smoke rolls into the hallway. Kieron is pushed into the cloud and the door bangs shut behind him. Through the haze of smoke he can see two pairs of eyes staring at him. There's a young guy sitting on a bed and a man standing in the middle of the cell.

Kieron is twenty-nine years old, a former journalist at *The Times* newspaper in London. He's tall, broad-shouldered and wears black rectangular glasses. He didn't join the ship to protest against Arctic oil drilling, though he's sympathetic to the campaign. Instead, he wanted to make a film about people willing to risk jail for a cause. But now he finds himself in prison.

Silence.

Kieron draws a breath but the smoke catches the back of his throat and he coughs. The man takes a step forward,

Kieron takes a step back. The guy is wearing tracksuit bottoms and a white vest, his arms are thick and hard, he has close-cropped hair, a flat nose, a heavy gold chain nestling in dense chest hair. He says something in Russian, Kieron raises his shoulders, a shrug that says he doesn't understand.

'You no Russian?'

Kieron hugs the mattress to his chest. 'No.'

'What is your name?'

'I'm ... I'm Kieron.'

'Kay-roon?'

'Kieron. Keer-an.'

'I am Ivan.'

'Ivan.'

'Yes, Ivan. Where you from?'

'London.'

'London.' The guy nods. 'Cool, cool.'

Kieron shuffles backwards.

'Why you here?'

'I'm a journalist. I was on a ship. You know Greenpeace? I was on ... '

And in one movement the man's face lights up, he spins around and lurches towards the window before shouting into the night, but the only word Kieron recognises is 'Greenpeace'.

What the hell is he doing? Christ, thinks Kieron, he's boasting to his friends that he's got a foreign cellmate, that's what he's doing. Oh Jesus. He's telling them he's got one of the foreigners, he's telling them what he's going to do to me.

There are ropes hanging from the window, manic shouting from the other cells, banging from the walls, the ceiling, the floor. Meanwhile the other guy, the one who hasn't spoken, is just sitting there on his bed, staring at Kieron with intense focused interest.

Ivan spins away from the window. Kieron steps back, presses his back against the inside of the door. This is it, he thinks. It's happening now. He's coming now, he's lifting a fist, okay here he comes, oh Jesus he's massive, he's … he's holding out a hand. Right. Okay. He wants to shake my hand.

Kieron hugs his bedding with one arm and extends a hand. The Russian pumps it hard.

'What is your crime?'

'Piracy. That's what they say.'

'Pirate. Yes, very cool. How old?'

'How old am I? Why do you … I'm twenty-nine.'

'London, yes?'

'Yeah, London.'

Charge, name, age, location. The Russian wants to know it all, then he pulls an exercise book from the shelf and writes it all down before returning to the window and shouting it all out into the night. And Kieron's thinking, is this guy FSB? I mean, if he is then it's pretty blatant.

In the cell next door, Phil Ball is looking at a skeletal figure in a dirty, crumpled shell suit. Three weeks ago Phil was enjoying the late English summer at his home in the Oxfordshire countryside. He was the last recruit to the crew after a late dropout. He's forty-two years old and has three young children. Now he's standing before a man with a shaved scalp and eyes sunk in deep, dark sockets. The door goes *bang* behind him, and instantly the man wants to know Phil's full name, date of birth, his home town and which article of the criminal code he's charged with. And Phil's thinking, you're a prisoner, what the hell has it got to do with you?

73

But then the guy crosses the cell and returns with a list of names and cell numbers, and some of the activists are already on there.

Next door, Ivan takes Kieron's bedding from him, throws a sheet over a bunk and tucks in the corners. He crouches down in the corner, pours water, opens tins, shovels powder then turns around to proffer a cup of tea and a biscuit. Finally the other guy is ready to introduce himself. He stands up, holds out a hand and says his name is Stepan. He's just a kid, shy, possibly not dangerous but Kieron's taking nothing on sight here. He's in full fight or flight mode, alert to every bang and thump, his wide eyes taking in the details of the cell – the peeling walls painted gooseberry green, the rusting bunks, a tiny barred window, a cubicle of rotting wood around the toilet, bags of food on a shelf on the wall, and ropes running out through the bars. The cell stinks of stale cigarette smoke and wet laundry. There's a loud bang on the wall. Stepan turns away and tugs on one of the ropes, like he's fishing and he's hooked a catch. Ivan darts over and pulls a sock through the window, plunges his hand into it and extracts a small piece of rolled-up paper. He unfolds it, looks down, reads it then offers the scrap to Kieron.

'For you.'

'For *me*?'

'Uh huh.'

Kieron plucks it from his fingers and holds it up. It's a written note, the size of a packet of cigarettes.

Hey man, I just heard about this thing called the road. It's like an email system but on ropes. Stay strong, we'll get through this. Po-Paul.

Next door, Phil is being handed a pen and a square of paper, and with wide eyes and a form of sign language involving rotating arms and energised pointing, the Russian guy explains that Phil can write to his friends. A minute later in cell 308 there's a bang on the wall. Stepan pulls in the rope again and extracts another piece of paper from the sock. He unfolds it and hands it to Kieron.

I'm not quite sure how this system works but I thought I'd try it out. Send me a message back if you get this. Phil.

Kieron asks Ivan for a piece of paper and a pen, and scribbles a message.

'What you say?' asks Ivan.

'Just a note to my friend. Some self-help bullshit, basically.'

Phil, we can get through this, we've got to stay tough, we've got to be strong for each other. Kieron.

Ivan rolls it up, drops it into the sock and bangs on the wall. The sock flies out of the window and a minute later Phil reads the note. He folds it, drops it into his pocket and glances up at his cellmate. The man says his name is Leonid. His lower gum is lined with a sparse row of teeth that look like broken fence posts. With more pointing and gesticulations Leonid explains how the rope system works, how all of the windows are connected throughout the night. It's a grid consisting of horizontal and vertical ropes across the outside walls.

'Prison internet,' he says.

On a corridor on the same floor, the Russian campaigner Roman Dolgov is holding a mattress, a blanket, a bowl and a spoon. The guard leans forward and pushes the door. Roman takes in the bright green walls, two bunk beds and a small barred window, then a head snaps up and a kid screams at him from a bunk. 'What are you doing? Get the fuck out of here!'

Roman flinches. The guard shouts, 'I haven't got a choice, okay. We're full up.'

'But you *know* I don't have anybody in here with me. This is a woollen cell, you know that!'

'I don't fucking care, he's coming in with you. It's one in the morning, I haven't got anything else. Deal with it, arsehole.'

Roman hugs his mattress and shuffles forward. He's a large man with a thick mane of swept-back hair and a long luxuriant beard. His voice is soft, his demeanour gentle. The door slams behind him. The kid slumps down on his bunk in a sulk.

'I'm sorry,' says Roman.

The kid scowls and stares at the wall. 'This is a woollen cell.'

'I'm sorry, I don't know what that means.'

'It means I should be alone.'

'I'd leave if I could.'

'You should be in a people's cell.'

'I'm sorry, I don't know the difference.'

The kid spins around and sits up. He's in his late teens, a bony face, wispy stubble on his chin, big blue eyes. 'There are people's cells, okay. Normal cells for normal prisoners. Thieves like you. And then there are woollen cells, cells like this. This is a woollen cell.'

Slowly, carefully, Roman steps forward and lowers himself onto the edge of the bed opposite.

'And what does that mean, to be in a woollen cell?'

'What does it mean? It means I'm way down the hierarchy. Waaaaay down, brother. But don't go thinking I'm *obizhenny*. I may be a piece of shit in this place, but I'm not that low.'

'I'm sorry, I—'

'*Obizhenny*.' It means 'morally injured'. 'You know, like poofs and perverts. This is a woollen cell, but I'm not *obizhenny*. When you're woollen you're on your own, that's the deal. But now you're in here and that means you're woollen too.'

'Is that bad?'

The kid throws his head back and laughs. 'Believe me, brother. You really don't want to be here. You'll have big problems later, when you get to the labour camp.'

'Why?'

'You're in a woollen cell! Aren't you listening? A cell for collaborators, for people who grass up the thieves. I'm a woollen guy. Couple of dealers wasn't it. Dealers like me. And an addict. I gave them up to the cops so the bosses got me put in here. They run this place, they got me demoted and now I'm isolated from prison society. And if you're in here then so are you. You want to watch it, mate. While you're in here you're only one step away from the bottom of the bottom, in with the poofs.'

Roman lays his bedding to his side on the bunk. 'Well … okay, but what should I do?'

'Don't panic. Not yet. The guards threw you in here, you had no choice.'

Roman nods. 'Exactly.'

The kid leans forward. 'Okay, listen to me. Here's some advice for you, brother. Think ten times before saying anything in this place. In prison, I mean.'

'Right, okay.'

'I'm serious. People can ask provocative questions around here. An example. Someone asks you if you're married, say yes, but if they ask you if you have oral sex with your wife then watch out. You say yes, they can say you're a pervert and make you *obizhenny* once and for all.'

'So … so what should I do?'

'What should you do?' The kid shakes his head, like it's obvious. 'Don't play tough guy. Don't boast. And keep your trap shut about your sex life.'

'But if it's a mistake that I'm in this cell, what can I do?'

'Break me out of here. So you're not tainted. If I leave, this place won't be woollen any more. Break me out.'

'Well how do I do that? I mean, I've only just arrived here. Am I supposed to fight you? Is that what you mean? I don't want to fight you. I don't even know you.'

On the second floor, Camila and Alex are alone in their cells, lying on their bunks. The air is filled with screaming and banging, the sound of locks turning and doors slamming. Somebody is thumping incessantly on Camila's wall, she shouts back, begging them to leave her alone. Alex has given up on sleeping, the wails of the other prisoners are filling her ears, the mattress is nothing more than a thin layer of lumps and gaps. The whole place is shaking, the Russians are screaming, alarms are going off, and she can still hear the dogs barking, vans pulling up. It's the other activists, more of her friends from the *Arctic Sunrise* being brought to jail from the courthouse. The prison is vibrating through the pipes, people are stomping on the floor above her. They're lunatics, she thinks. This place is crazy. She can't sleep. She's cold. She has a purple ski jacket on and a blanket wrapped around her, but she's still freezing.

Upstairs Kieron is lying on his back, a bright light flickering over his head. He doesn't have any warm clothes and his teeth are chattering. He manages to sleep for a while then he wakes up shivering, sleeps again, wakes up. Ivan is squatting in the corner of the cell cooking food while Stepan is working the ropes. Ivan stands up and brings a bowl over to Kieron.

'Here, breakfast. For you, my friend.'

Kieron attempts a smile, nods his thanks and takes the bowl. Ivan doesn't move, he's standing there waiting for Kieron to try it. Kieron raises the spoon to his mouth, but before it's touched his lips he can smell it. It's disgusting. It looks like porridge but it smells of burning rubber. Kieron gulps and slides the spoon between his lips and swallows. Ivan smiles. And Kieron's thinking, just eat this shit, don't upset your cellmates, get through the night.

In the cell next door Phil wakes with a jolt. It takes him a moment to remember where he is. It's dark outside. He falls asleep again, wakes up. He hugs the sheet, drifts off, then wakes to the sound of metal on metal. The hatch in the door is open, a guard is shouting something and holding a bowl. Phil stands up and takes it. He looks down at a grey paste that smells of chemicals. The hatch closes.

It's Friday morning, 6 a.m., and their first full day in SIZO-1 is about to begin.

EIGHT

A guard and a female interpreter are stood over Alex Harris. She gets to her feet. The guard peers over Alex's shoulder and looks around the cell then says something. The woman translates, asking Alex if she has any questions.

'I just want to know, when can I speak to my family?'

Guard and translator converse.

'I don't know.'

'Please. When can I speak to my family?'

More words are exchanged, the woman turns back to Alex.

'You have to put an application in to the investigators.'

'I have to do *what?*'

'Yes, a written formal application.'

'To make a phone call?'

'Yes.'

'But … but surely that's a basic human right, to make a phone call. I've just been locked up and … '

'I'm sorry, there's nothing I can do.'

'Okay, so I have to write this application. How long does it take?'

A long and involved consultation between translator and guard ends only when he shrugs his shoulders and fiddles with his belt buckle. The woman says, 'A few weeks, maybe a month.' And with that, the guard coughs, steps backwards, reaches out and pulls the door closed.

Two thousand four hundred miles away, Cliff Harris and his wife Lin are staring at the television. Every hour the news channel runs pictures of vans leaving the Investigative Committee headquarters, and they know their daughter is in one of them. The family is in a state of shock. They don't know how Alex will possibly cope with the ordeal she's undergoing. 'You have to start from the fact that Alex has always been a very sensitive, caring child,' says Cliff.

But Alex isn't feeling very caring right now, instead she's burning with indignation, chewing her lip with anger and staring at the door, ready to jump down the throat of the next person who opens it. Hours pass before it swings open again and two guards stand before her.

'I want my telephone call,' she shouts at them.

'*Gulyat.*'

'What? What the fuck does that mean? I want my … ' She sits up and makes a phone shape with her thumb and little finger and holds it to her head. 'I want to call my family.'

'*Gulyat.*' One of the guards makes his fingers do a little walk. 'Exercise.'

Alex's heart jumps. At last, this is where she'll see her friends. She assumes it's going to be a big yard like in the movies, with prisoners in orange overalls playing basketball. It's going to be good to see the others again, to hold them and share stories about this place.

She pulls on her purple ski jacket. The guards lead her down hallways, through doors, down a staircase and outside into the open air. They're walking her towards a concrete building with a long line of doors. When they get there they open one and push her inside. With a scraping sound the door closes and a key turns in the lock.

It's dark. She's in a box. It's two metres by three metres. She looks up. The roof is made of chicken wire and crumbling asbestos tiles, and through the mesh she can see a guard cradling a rifle, looking down at her from a bridge. The floor is covered in spit and cigarette butts. She kicks out, spins around and hugs herself. Then she buries her face in her ski jacket.

In all her life, she has never felt so alone.

Alex Harris studied marketing at university and always assumed she would end up at an advertising agency. On her placement year she worked at Bosch power tools. After graduating she saved up money in Abu Dhabi – 'not my kind of place' – and hit South America. And that was where she fell in love with nature. She was in the Amazon, on a canoe, it was idyllic, birds flying above her in the dense jungle canopy. Then she saw the oil pipelines. And she thought, why are they there? Why the hell are they pumping oil through the Amazon jungle? That's crazy.

After that she went to the Galápagos Islands. She dived with sharks and swam with seals and turtles. When the *Deepwater Horizon* oil platform blew up in the Gulf of Mexico, she was in Australia. She watched the TV footage of oiled coastlines and felt real, visceral anger. She wanted to do something. She bombarded her parents and friends with petitions. Then she signed up to volunteer at Greenpeace.

Two years later she boarded the *Sunrise* and sailed for the *Prirazlomnaya* as a full-time campaigner. And now she's locked in a stinking box in a prison in the Russian Arctic.

'Is anybody there?'

She jolts and looks up. It's a familiar voice coming over the wall from the box next door.

'Cami? Is that you?'

'Alex!'

'Camila! Are you okay?'

'I'm okay, are you okay?'

'Not really.'

'What's your cell like?'

'Oh my God, it's horrible.'

'I'm scared, Alex.'

'The guards are arseholes.'

'Did you eat the breakfast?'

'Are you kidding me. It smelt of cold sick.'

'We're gonna starve in this place.'

'Are you next to me?'

'I don't know.'

'Which cell are you in?'

'Two-sixteen.'

'I'm in two-fifteen! We're next to each other!'

'Maybe we can talk.'

And right there, on that first morning in SIZO-1, they devise a code. They agree to tap spoons on the pipe that connects their cells. One tap is 'A', two taps 'B', three for 'C', twenty-six taps for 'Z'. Five taps mean you want to speak, three means okay, go ahead. And an hour later, when they're back in their cells, they have their first coded conversation. Then at 10 p.m., when the stamping starts again, when she hears the screams of the other prisoners, Alex taps to Camila and gets a tap back. And that's how they reassure each other. That's how they know they'll survive the second night.

The women have heard nothing of the *doroga* – the road. They have no cellmates to initiate them into the rituals of this place. Meanwhile the men exchange dozens of emails with each other. Kieron discovers he can get a message to almost any cell in the prison. But mostly he exchanges letters with Phil. They discuss rumours they're hearing from the others, how some of the activists are saying this will be over in a few days or a few weeks. But even then Kieron's thinking, no, I can't cope with this, I can't do a few weeks in here.

The hours stretch out in a monotonous routine that takes just a single weekend to become familiar. Porridge arrives at six. You eat it then go back to sleep. At eight the guards come into your cell, you put your arms in the air and they frisk you. Then the guards take you out of the cell and stand you in the corridor with your hands against the wall. A guard wipes his palms down your trousers while another guard searches your cell. When the search is over you're pushed back into your cell and the rest of the morning is yours until the guard comes back and says '*Gulyat*' and you say '*Da*'. Then you put your coat on and off you go to the exercise box.

Phil walks there in his boots without laces, shuffling beside a guard, with the camera card under his heel. Then he comes back and spends the afternoon lying on his bunk or staring through the window at the basketball court below. All weekend he sees just one person down there. It's a child. A young boy. The guards say it's the son of one of the women prisoners. He's out there, bouncing a ball, alone.

In the town of Irvington on the outskirts of New York City, 73-year-old Pavel Litvinov is sitting at the desk in his study, scrolling through Facebook, when he sees a status update saying his son has been jailed. It's nearly forty years since Pavel left Moscow on a train to Vienna with his wife and children, expelled by the KGB. And now Dima is back in Russia, and in prison.

Pavel's life in America has been comfortable. He thought the dramas of his youth under the Soviet system were behind him. After his expulsion in 1973 he took a job as a physics and mathematics teacher at a prep school, eventually retiring seven years ago. And he still looks like a teacher. He has a bald pate, wears check shirts and a cloth cap and speaks English with a strong Russian accent, despite living more than half his life in the USA.

Now Facebook is telling him that Dima just appeared before the Leninsky Court in Murmansk. The case was postponed from Thursday but today it was quickly dispatched. The judge told Dima he would be jailed for at least two months while an investigation into piracy continues.

'And I was devastated. I was just so scared and so upset. I already knew Dima was there on the ship but I didn't know the details, I didn't know how far he'd got. I was in deep fear and depression. When Dima was jailed my reaction was almost irrational. I gradually pulled myself together, but initially my reaction was fear. I felt I'd taken Dima and his sister out of that country and it would never touch them again in a scary way. So it was just terrible. I called Human Rights Watch and Amnesty International, then I called my local congressman. I was probably more scared than anything, because I thought that at this time the Russian regime was going back to the Soviet ways, and Dima is going to be in a position where this rollercoaster will come down and crush him.'

Pavel doesn't know it, but last night a documentary was broadcast on the Russian national television station NTV, which is owned by Gazprom. Called *Under the Green Roof*, it made a series of outrageous claims against Greenpeace. The same station previously broadcast a controversial programme smearing the Pussy Riot protesters. Now Pavel's son is getting the same treatment.

The documentary claimed Greenpeace is a tool of the American government, that the organisation has been bribed by Western oil companies to ignore their polluting activities, that Greenpeace stayed silent during the *Deepwater Horizon* disaster and never criticises Exxon – the American oil giant. The programme claimed Greenpeace forged footage of seals and kangaroos being abused, and is banned in Canada because it's labelled an extremist organisation. The documentary ended with the claim that Greenpeace is controlled by international financiers and banking clans.

After Dima's court appearance comes Sini Saarela – the Finnish climber who scaled the *Prirazlomnaya*. She stands in the cage and addresses the judge. 'I am an honest person,' she says. 'I am always ready to take responsibility for what I have done. I am not a pirate. Drilling for oil in ice is a tremendous threat to the environment in Russia and across the Arctic.' But the judge is unmoved. She gets two months. Next comes Frank, then five others. They're all told they're going to jail.

The activists are loaded into an *avtozak* transport van outside the courthouse. It's late, they can just about make out the bleak Soviet architecture as they're driven through the city. They hear gates opening, people screaming and shouting in Russian. The van stops and waits. It moves again. Then it stops and the doors are flung open.

'Go! Get out get out get out! Welcome to your new home! A step to the left is considered an escape attempt, a step to the right is considered an escape attempt. Move, get out get out get out!'

Dima throws his huge pink bag over his shoulder, steps out and looks up. And this is so incredibly familiar to him. He knows this. He knows this order. *A step to the right or a step to the left is considered an escape attempt.* He's read this in so many books; he's read this so many times in Solzhenitsyn and the other prison camp novels he was brought up on. This feels natural. Like fate. Unavoidable. At some point he had to be in prison in the country of his birth. There's a Russian saying – *ot turmy da ot sumy ne zarekaysya.* 'Don't swear you'll never go to prison.' But Dima knows too much about places like this, there's too much in his blood. He was always going to end up here.

'Welcome to your new home.'

Yes, thinks Dima. I'm coming home.

NINE

Many a man thinks he's shifted the course of human history from an English public house, but Maxim Litvinov actually did.

The year was 1907 and the place was a dockers' pub in the East End of London. The Fifth Congress of the Russian Social Democratic Labour Party – one of whose leaders was Vladimir Lenin – was being held in a nearby church. Also there was Lev Bronshtein, who'd arrived in England after escaping from a Siberian prison and travelling 400 miles across frozen tundra on a sleigh pulled by reindeer. He was known to his comrades as Trotsky.

Maxim Litvinov was staying in a sixpence-a-night doss-house with his friend, the Georgian revolutionary Josef Dzhugashvili. By day they attended the Congress together, listening to speeches and debating the party's strategy for overthrowing the Russian tsar and instigating a communist dictatorship in their homeland. By night they drank.

Dzhugashvili had a reputation for thinking with his fists. One evening they stumbled into a pub frequented by London's notoriously belligerent dock workers and soon the Georgian was embroiled in an argument with the locals. It soon escalated, Dzhugashvili was surrounded, and although the details of that moment are lost to history, one can imagine his life flashing before his eyes. That was when Litvinov waded in and dragged his hot-tempered friend from the pub. They escaped down a side street.[10]

Maxim Litvinov was Dima's great-grandfather. Josef Dzhugashvili survived that night to become one of history's greatest mass murderers. He is known to the world as Stalin.

Dzhugashvili was largely unknown to his comrades in London, but Litvinov was already a notorious revolutionary. He'd been arrested by the *Okhrana* – the tsar's secret police – and jailed in Lukyanovskaya prison, from where he escaped, taking ten inmates with him.[11] He slipped over one border after another until he reached Switzerland. Later he became an arms dealer for the Bolshevik wing of the party, procuring rifles and smuggling them into Russia through Finland.[12]

A year after dragging his friend from that London pub, Litvinov was arrested[13] in France carrying a pocketful of 500-rouble notes that had been stolen in a notorious bank robbery masterminded by Stalin.[14] He was deported, returned to London and married an English woman, Ivy Low. Litvinov was still in England when Lenin and Trotsky seized power in the Bolshevik Revolution of 1917. He was arrested by the London government in 1918 but freed in exchange for a British diplomat who was accused of complicity in an assassination attempt on Lenin.[15] Litvinov returned to Russia and began a career as a senior Soviet diplomat.

In 1930 his old friend Stalin – by now supreme leader of the Soviet Union – appointed him Commissar for Foreign Affairs. Litvinov – real name Meir Wallach-Finkelstein[16] – worked to normalise relations with Britain and France, and persuaded America's President Roosevelt to officially recognise the Communist government in Moscow. And all the time he was a senior figure in a regime that was purging, starving and shooting millions of its own people.

In April 1933 Litvinov's face graced the front cover of *Time* magazine.[17] The Nazi government in Germany derided his Jewish ancestry, with Berlin radio referring to him contemptuously as 'Finkelstein-Litvinov'.[18] When Stalin resolved to sign a pact with Hitler to invade and divide Poland, Litvinov's Jewish roots presented an awkward impediment. His office was surrounded by troops from the NKVD[19] – the secret police organisation that succeeded the tsar's *Okhrana*. A delegation led by Vyacheslav Molotov told Litvinov he was fired.[20] Four months later Molotov signed the pact with Hitler, Poland was invaded and the world went to war. Hitler later said, 'Litvinov's replacement was decisive.'[21] Asked why he had been replaced by Molotov, Litvinov said, 'Do you really think that I was the right person to sign a treaty with Hitler?'[22]

When, two years later, Germany turned on the Soviet Union, Litvinov was rehabilitated and appointed ambassador to the United States. On New Year's Eve 1951 he died of a heart attack, aged seventy-five.[23] Molotov later said Litvinov was 'utterly hostile to us ... He deserved the highest measure of punishment at the hands of the proletariat. Every punishment.' He said Dima's great-grandfather 'remained among the living only by chance'.[24] Decades after his death, Litvinov's daughter claimed Stalin once told Maxim he was only spared because, 'I haven't forgotten that time in London.'[25]

Lev Kopelev was also a Bolshevik. In the 1930s he worked as a journalist, witnessing the horrors of the Ukrainian famine caused by the forced grain requisitioning ordered by the government of which Maxim Litvinov was a part.[26] When the Germans invaded in 1941, Kopelev volunteered for the Red Army, serving as a propaganda officer.[27] He was one of millions of Soviet soldiers who rolled into Germany near the war's end. And it was here, in East Prussia, that he witnessed atrocities committed by his nation against the defeated German civilian population.

Kopelev was deeply troubled and felt unable to remain silent. He spoke out publicly, denouncing the conduct of the Soviet armies in Germany. He was promptly arrested[28] and sentenced to ten years' imprisonment for the 'propagation of bourgeois humanism, sympathy with the enemy and undermining the troops' political–ethical morale.'[29] In the gulag he met and befriended Aleksandr Solzhenitsyn. When Solzhenitsyn came to write his novel *The First Circle,* he based the character Rubin on Kopelev.[30] When Kopelev was released, he approached Russia's leading literary journal and urged it to publish Solzhenitsyn's *One Day in the Life of Ivan Denisovich.*[31] It was a seminal moment in the Soviet dissident movement.

Lev Kopelev was Dima's grandfather.

In 1968 Lev was expelled from the Communist Party[32] for lending his voice to protests against the persecution of other dissidents. He also spoke out against the Soviet invasion of Czechoslovakia.[33] It was a cause for which Pavel Litvinov – Kopelev's son-in-law, Maxim's grandson – was also prepared to pay a heavy price.

As a high school student, and even during his early years at university, Pavel was a devoted member of the Young Communist League. But by the end of his time as a student his commitment to Marxism had collapsed. Ideology had clashed with reality and he viewed Soviet society with 'cynical indifference'.[34]

Pavel became a physics teacher. He befriended a group of intellectual anti-Soviet writers and worked for the release of political prisoners, hungrily consuming *samizdat* literature – banned publications which now included the works of Solzhenitsyn.[35] In 1967 he was pulled into KGB headquarters and warned he was risking arrest and imprisonment for supporting dissidents, but Pavel made a verbatim record of the interrogation.[36] It was published in the *International Herald Tribune* and four months later he received a telegram from 'a group of friends representing no organisation' who 'support your statement, admire your courage, think of you and will help in any way possible'. The letter was signed by Yehudi Menuhin, W.H. Auden, Henry Moore, Bertrand Russell, J.B. Priestley, Paul Scofield, Sonia Brownell (who signed as 'Mrs George Orwell'), Cecil Day-Lewis and the legendary Russian composer Igor Stravinsky.[37]

'They transmitted the letter on the BBC in Russian,' says Pavel. 'And the BBC called who they could and asked them why they signed this telegram. So they called Stravinsky, and Stravinsky was already a very old man. And he said, "We have to support Litvinov because my teacher, the composer Nikolai Rimsky-Korsakov, also suffered from Russian censorship." And I started to cry. It was so touching. Rimsky-Korsakov was history from one hundred years earlier, he was played on every radio station in Russia, and suddenly through Stravinsky I connected with Rimsky-Korsakov.'

The following spring Pavel married Lev Kopelev's daughter Maya. The old man was an inspiration to Pavel, an example of how someone could live outside the system, in his heart at least.

On 21 August 1968, Soviet tanks entered Prague to suppress a nascent move by its reformist government, led by Alexander Dubček, to implement 'socialism with a human face' – at that time, by definition, a break with the Soviet Union. Demonstrations broke out across the world, tens of thousands took to the streets to protest the invasion, in Prague itself many demonstrators were shot by Soviet troops. But it was inconceivable that there would be protests in the USSR itself.

Until Pavel Litvinov and seven of his friends resolved to act.

On the evening of the twenty-fourth they went to the Kopelevs' apartment for a party, fully aware it would be their last night of freedom. 'We knew we were going to prison for years,' Pavel remembers. 'To a labour camp.' The following day they would certainly be jailed, but they were ready. The famed singer Aleksandr Galich was at the party, and at one point he began to sing a protest song that was popular decades earlier among opponents of Tsar Nicholas I.

Can you come to the square?
Dare you come to the square
When that hour strikes?

Pavel listened silently, but inside he felt the baton of resistance being passed. He almost announced what would happen the next day, but he remained silent, not because he feared betrayal but because some of the older dissidents gathered in the flat might insist on joining him. He doubted they could survive the retribution of the Soviet state.[38]

The next morning, a Sunday, Pavel and the others walked towards Red Square. They were being followed, and they knew it. When they reached the *Lobnoye Mesto* – the Place of Skulls – they sat down and unfurled a Czechoslovak flag. Officers from the KGB descended on the group. The protesters had no more than a moment to lift their hand-painted banners.

> *Shame to the occupiers!*
> *For your freedom and ours!*
> *Long live free and independent Czechoslovakia!*

The KGB rained blows down on their heads, shouting, 'These are all dirty Jews!' and 'Beat the anti-Soviets!'[39]

'They hit Viktor Fainberg,' Pavel remembers. 'He was sitting next to me and broke four of his teeth. They beat me with a bag. It felt like it was full of bricks but I think it was books. The adrenaline was so high that I didn't notice much, but they beat me very hard. Later I touched the top of my head and it hurt like hell.'

The group was bundled into unmarked cars. Before they could be driven away to the nearby KGB headquarters a policeman's whistle blew, and from the Kremlin's Spassky Gate there came a line of black cars that drove right past Pavel and his friends. Later they would learn that Alexander Dubček was inside one of those cars. The leader of the Czechoslovak rebels was now himself a prisoner.[40]

Pavel was charged and jailed, his trial set for six weeks' time. It was said that the Soviet leader Leonid Brezhnev and his KGB chief Yuri Andropov were incandescent that such a protest had occurred at the very centre of Soviet power,[41] in front of the captured Dubček no less, and it was widely known that the verdict and sentences in the coming trial had been decided long before the proceedings even began. In his book *Lenin's Tomb* – a Pulitzer Prize-winning account of the fall of the Soviet empire – the journalist David Remnick would later write that the protest in Red Square 'struck one of the first blows against the regime'.

The trial came and Pavel was convicted. But he wasn't jailed. Instead he was sentenced to exile in Siberia. The regime had decided they didn't need another martyred dissident, especially one with a famous name. Pavel packed a suitcase and left Moscow with Maya and his six-year-old son, Dima.

For five years the Litvinov family lived in exile in Usugli, a village lost in the vast Siberian *taiga*, where the forests go on for hundreds of miles. If you walked away from that village and got lost, you'd die. That's what Dima remembers of his childhood in exile. Nature was all around them, and it was so much more powerful. It was in control.

In December 1973 the family was back in Moscow. The KGB pulled Pavel in again and made him an offer he couldn't refuse. Either leave the USSR permanently or be sent to the gulag – a Soviet labour camp.

Pavel feared that if the family stayed in the USSR, Dima himself would eventually be jailed. 'My wife Maya and me were thinking something will happen to Dima because it was clear that if I didn't emigrate then I would be arrested again. And it was clear that a teenager in a dissident family at that time cannot help but become rebellious and probably will go to prison. In a way it sounds strange but the Soviet state in my time under Brezhnev tried to respect the law. Not human rights but the law, more than it does now. It's much more arbitrary now. The Soviet regime in my time was kind of more stable, they knew what they were doing, they were prepared, they were moving away from Stalin and his regime, they tried to demonstrate that everything was legal. Putin's regime is more improvising. That's why it is more scary.'

Three months later Pavel Litvinov left Moscow on a train to Vienna with his wife, his son Dima and their infant daughter Lara.

The family moved to the United States. Dima was twelve years old. He went to high school then studied anthropology at college. For his thesis he moved to Ecuador to live with highland tribes, and there he met a Swedish woman, Anitta. They married in Guatemala. When a son was born, they named him Lev.

Dima's grandfather Lev Kopelev was always one of the most important people in his life, so Dima saw his son's birth as a continuation of the line. It was having kids that brought him into Greenpeace. He was reading *Newsweek* at home with his boy, holding this baby in his arms, reading about climate change and environmental destruction. And Dima thought, it's not enough to just know about it, you have to do something.

Now, forty-five years after his father's arrest and imprisonment by the KGB, sixty-eight years after his grandfather's imprisonment by the NKVD, and 112 years after his great-grandfather's imprisonment by the *Okhrana*, Dima Litvinov is in the reception yard of a Russian prison, the fourth generation of his family to be jailed by the Russian secret police for his political beliefs. Only now they call it the FSB.

'When Dima was with me in exile there was no reason to worry about him,' says Pavel. 'It was just our life, his mother and me, and later our daughter Lara when she was born.

There was no reason specifically to worry. I was the only one technically punished. We had a hard life and it was tough. It was extremely cold. I almost died from a bad case of pneumonia when I was working in the mines. There was a doctor who became my friend; he worried about me and they sent me to hospital. They didn't want me to die on their watch. It would be bad publicity. But that was only happening to me, so there was no fear at all. But here I was, already a pretty aged man living in comfortable America. And suddenly my son gets sucked back into that life.'

TEN

'Okay, hands against the wall! Hands against the wall, bags on the ground!'

There's a five-metre-high gate behind them, and in front of them is another set of identical gates with tight rolls of barbed wire over the top. A blinding spotlight is trained on them, dogs are barking all around.

'Pick up your bags! Turn left! Keep one hand behind your back. Move!'

Ropes are hanging vertically and horizontally from the brightly lit windows, socks are being pulled across the wall. And Dima knows it's the *doroga* – the road. He looks up, and the feeling he has when he sees it is … is *joy*. He knows what all this is and he's finally going to experience it. He knows it from books, from family stories, from songs sung at home. It's in his blood. And now he's here, he's actually going to live it. But the air is filled with screaming and thumping. He glances at the faces of his friends. A guard pokes him in the back.

'Move!'

In an instant the joy deflates. No, this isn't a book. This is real.

They're each issued the standard bedroll, an aluminium bowl, aluminium spoon, aluminium mug. Then they're taken out into a long broad corridor with rows of pitted metal doors on either side. The hallway echoes with clinking keys, shouted orders, the cries of the other prisoners. One of the hatches ahead is open, Frank can see a face squeezed through the gap, and as he gets closer he can see it's Kieron. Their eyes meet. He doesn't look good. Wide eyes, messy hair.

Frank is stopped outside a door. The guard pulls it open and pushes him inside. It smells of cigarettes and damp. In front of him two men are pulling a rope through the window, and Frank thinks, Christ almighty, they're getting the drugs in, I'll keep well out of that, I'll just keep myself to myself.

But a moment later the men have dropped the rope and are questioning him.

'Name? Birthday? Where you born?'

Frank bites his lip. He considers ignoring them but he thinks better of it. He gives them his full name and date of birth.

'Where you born? What crime?'

Frank tells them. He sees one of the guys writing it all down, then the Russian drops a scrap of paper into a sock and it disappears out the window. And Frank thinks, fuck, identity theft! What an idiot! I've been here one minute and I get suckered. They're gonna rob my bank account. I'm stuck in prison, this goes to the bosses and they sell it to some guy on the outside. Unbelievable.

Dima is standing outside cell 306. He rubs his short salt and pepper hair, scratches his beard and pushes his round, steel-rimmed spectacles up his nose. The guard inserts a huge key into the lock. The door swings open. Dima steps inside, he puts down his pink bag and the door slams shut behind him.

And he thinks, yes, definitely a Solzhenitsyn moment.

Dima knows the protocol from the books. There are four beds, three inmates. The bottom beds are taken, he nods to his cellmates and throws the bag onto a top bunk, turns around and introduces himself.

'Litvinov, Dimitri. Born in sixty-two.'

'Vitaly.'

The other guy says, 'I am Alexei. Welcome.'

Then Vitaly says, 'What are they charging you with?'

'Piracy.'

Dima's cellmates stand silent for a moment before they both make incredulous little circles with their lips. 'Ooooohhh,' says Vitaly. 'We've been expecting you. Sit down sit down, be comfortable, my friend. Yes yes, we knew you were coming. Didn't know we'd have one of you in this cell, but we knew you were coming to SIZO-1. Some of your friends are already here. There was a memorandum from the *kotlovaya*, the boss cell, it said we should be positive and co-operative with you. In the criminal hierarchy you're pretty high up, you know. Because you're sufferers. You've suffered from an absolute injustice. Yes yes, we knew you were coming. We've known for a week. We knew before your judge found out.'

Down the corridor Frank is lying on his bunk, staring at the ceiling, pondering those words shouted at him through a cell door back at the courthouse on Thursday. *You've got some fucking questions to answer. It's your fault we're getting sent down.* Is all this his fault? He was in charge of the action, that much is true, but surely nobody could have known the Russians would overreact like this? He knows who said it, and he knows some of the others will be thinking the same thing. Even if he gets out of here, he's still going to get shit from them. But then, maybe they're right. Or maybe not. Jesus, who knows?

Then Frank gets tapped on the shoulder and he's handed a little scrap of paper. It's from Dima, it says, *This place is fucking cool man, my cellmates are fucking great, I could stay here for months!* Then Phil sends him a note saying, *Frank old bean, nice of you to join us!* Then another one from Phil. *Beware the soup, here be dragons!*

Next door, Vitaly grabs Dima's hand and sits him down on the bed. And in furious excited Russian he launches into a crash course in surviving Murmansk SIZO-1.

'It's probably kind of weird for you, and scary to be here. But listen, Dima, people *live* here.' He's in his thirties with dark hair, yellow teeth, maybe Uzbek roots, possibly Tartar. He has light coffee-coloured skin with an alcoholic face, but he's been in prison so long the booze has drained from his cheeks. His skin is dry and shot with red hairlines from burst capillaries. 'This is not the end of the world, people live here just fine, and you will be fine here too. How many of you are there?'

'Thirty.'

'Thirty, right. Okay, well you can talk to them on the *doroga*, it goes all night, it goes to all the cells, you can send a message, there's no problem. We have another big group here, seventeen men. A gang. They shot up a nightclub. They stay as a gang here by communicating on the *doroga*.' He points at the wall. 'You see these shelves? We put all the stuff on there. Anything that's on the shelf you take, and anything you have that you want to share with us you put on the shelf. If there's something you don't want to share, keep it in your bag. If someone takes it from your bag they're a bitch and an arsehole, so nobody does that. So whatever you have, put it on the shelf. You're with us here now, we share everything and you should too.'

'Okay, cool, got it. What's mine is yours, what's yours is mine.'

'Exactly. Okay, what else? You have the morning inspection, and after they're done they say, "Do you have any complaints or questions?" And you'd better not say yes, because it doesn't matter what you say. If you say you have a problem, they'll give you a real problem. So just say nothing.'

There's a bang on the wall, Vitaly holds up a hand in apology, jumps up and pulls a sock into the cell. He unfolds the note.

'Aaaah, it's a *kursovaya*. It's a circular, sent from the boss cell. The normal letters are called *malyavas'* – 'deliveries' – 'but this one goes to all the new people, you and your friends.'

He hands it to Dima. It's written in prison slang – an ornate language that is both rough and formal.

The best of day and time to you, all arrestees! Here is hoping this note finds you in good health and strong of mood. Here is the deal. There is us and there is them, there are thieves and there are stars. The stars have stars on their shoulder plates, and these, dear friends, are the guards. Then there is us, we are the arrestees. We are the thieves. Now, the doroga *is most important, it keeps us as one, together, in solidarity. It is what keeps us alive. If there is anything you need, you will have it. All you need do is ask. You will not sell or buy things, no, you are expected to give. If you have something, you give it. If you need something, it will be given to you. If you want to be a part of the* doroga*, you are welcome to join our community of ropes, you will be supported, you will be given what you need. If however you are afraid to be a part of the road, we understand, and you will still be given support. But do not interfere with the* doroga*. If you interfere with the road then you will be punished, you will no longer be part of us, you will be one of them. You will no longer be a thief. You become a star.*

The note sets out other rules. Violence is absolutely not allowed. No arrestee is permitted to commit violence against another arrestee, if they do they will be punished, and they will be punished with violence. Only sanctioned violence is permitted, and it is for the *kotlovaya* – the boss cell – to determine if, when, how and against whom violent retribution is wrought. And you are not permitted to be rude. Hard cursing is not allowed against another prisoner. One is permitted to say, 'I hate this fucking shit,' but you can't say, 'Fuck you.' You will treat other arrestees with respect.

Dima finishes the note, shakes his head with incredulity and hands it back to Vitaly. The Russian consults his list of names and cell numbers, writes an address on the note, folds it then drops the *kursovaya* into the sock and bangs on the wall. It whips away, heading for another activist. Vitaly turns back to Dima.

'This is a black zone. There are black zones and red zones. That means there are things here that are not allowed but are still tolerated by the guards. Other prisons are red zones, that means nothing is tolerated. It's a much harder job for the guards in a red zone. The prisoners in those places are in for their fifth or sixth stretches, they've got ten-year terms, they don't give a shit. But this is a black zone. That's why the road is tolerated. They know it happens, just don't get caught.'

'I don't want any trouble.'

'Who does, Dimitri, who does? There are six walls here, three facing into the yard – the ones you saw when you arrived – and three facing out to the street. On each wall there is a boss cell. All the goods, the sugar and the cigarettes, everything, it all flows to and from that cell. The road operates on each wall, and each wall has its own *kotlovaya*, its own boss cell, its respected prisoner. So we have six bosses, responsible for maintaining order in our community.'

'The mafia.'

'No, no, Dimitri, please. We prefer to call them respected prisoners. You should too.'

'Right.'

'Now, the boss decides which prisoner goes into which cell. Of course he can't tell the guard to put this guy into cell three-zero-six, but what he can do is determine what category of cell some prisoners go to. They tell the guards and the guards co-operate with the *kotlovaya*. And there are basically four categories of cells. There are cells for the normal prisoners who participate in prison life – me, you, your friends – and we call them "people cells". The prisoners in those cells are the ones who get the respect, they're the ones who get decent treatment from the system, right.'

'Right.'

'Then below that are the *sherst* – it's the Russian word for 'wool' – 'the informers. If they demote you to *sherst* they have you put in a certain cell. The guards don't want any killings, right. They don't want any trouble. So when the boss says, "This guy, we want him in the *sherst* cell," the guards move him there.'

'Okay.'

'Then there are the cells for passive homosexuals.'

'Passive homosexuals?'

'In some ways we're more tolerant in here than on the outside. It's okay to be a fucker, but not a fuckee. It's not okay to give a blow job to a guy, but it's okay to get one. You can cum, no problem, but you can't put out. If you do, you're *petuch*. A passive homosexual, and that's bad. The passive homosexuals are a caste. They're the ones who clean the toilets, they do the shit work. Sometimes we make them wear dresses.'

'And they have their own cells?'

'Yes, that's right. *Petuch* cells. Let me see your bowl and spoon.' Vitaly gestures with his fingers. 'Come on, give them to me.'

Dima holds them out, the Russian snatches them and examines each in turn. 'Okay, good. You're not marked down as a poof. If they have you down as a poof they put the number "2" on your bowl and spoon. If you have a "2" on one of these that means the bosses have got you marked as *obizhenny.* Then you're demoted to a *petuch* cell. Okay, then below that there's another category. Former employees of law enforcement agencies. Cops. Prosecutors. There are lots of them in prison, there's a lot of crime that goes on in that sector of society. Bribery, murder, everything. And they end up here. They have their own cells as well. They keep themselves to themselves, otherwise they tend to get killed.'

Dima blows out his cheeks and whistles. Vitaly stands up and pulls an exercise book from the shelf. 'And this … ' He holds it reverently. 'This is the *domovaya*. Every cell has one, this is our house book. This needs to be kept religiously. This is where we keep the list of prisoners' names and their cell numbers for the *doroga*. The *domovaya* is very much a challenge for the regime, because we prisoners are not supposed to know what's going on beyond the walls of our cells. We're supposed to be in isolation.'

SIZO means 'isolator'.

Vitaly tells Dima that the *domovaya* allows the bosses on the wall to maintain discipline. So if somebody is a *sherst*, if he's sold out another prisoner and the bosses want to know where he's been transferred to, it's all in the book.

'And if somebody is abused by the guards, you want to know where they are so you can support them. It's very important that we maintain our community. As soon as somebody is put in your cell, you send a *kursovaya* to the whole prison saying there's been a change in my cell, such and such has moved in. His name is this, his crime is this, his date of birth is this, and that's all noted in the *domovaya* in each cell.'

'So you guys know where all my friends are?'

'Of course. And if a letter or a package passes through your cell on the way to another cell we will keep track of it, keep a record. Received and sent from this cell to that cell. Each cell is required to do that so you can compare it later. That way, if a package disappears along the way we can tell who lost it, what happened. Although that doesn't apply to the wet letters. Then we … '

'I'm sorry, *wet* letters?'

'Letters to the women's zone.'

107

Vitaly explains that he and his cellmate Alexei have girl-friends in the women's sector on the second floor. Lots of the prisoners are conducting relationships inside SIZO-1, though they've never met their lovers and they likely never will. 'Our love is as strong as anything you know. Those letters, our love letters, they have a different status on the road. Not the same status as normal business, where the rules are very strictly enforced.'

Dima flips through the *domovaya*. He looks up.

'Holy shit, you guys are pretty well organised in here.'

'You sound surprised. What else are we going to do? We have many days to fill, my friend.'

'Right.'

'Oh, and Dima, one more thing.'

'Sure.'

'We have a saying here. *Ne Ver' Ne Boysya Ne Prosi.* It is a fine motto. You can live your life by it. It tells you everything you need to know. It will help you survive.'

Ne Ver' Ne Boysya Ne Prosi. 'Don't Trust Don't Fear Don't Beg.'

'Don't trust anybody in a uniform,' says Vitaly. 'The more faith you put in the authorities, the more it hurts when they screw you over. To trust the police is to disrespect yourself. And don't fear because whatever you're scared of, you can't stop it happening. What will be will be. Your fear changes nothing, but it hurts you, so let it go. And don't beg because it never works. Nobody ever begged their way out of SIZO-1, so don't sacrifice your dignity on a false promise. There's no point being nice to the guards, the investigators, the prosecutor or the judge. Your pleading only makes them despise you more.'

ELEVEN

At SIZO-1 the policy is to hold prisoners accused of the most serious crimes in the same cells. Because piracy carries ten years minimum, most of the activists are held with Russians accused of killing or maiming their victims.

Frank's cellmates are Boris and Yuri. Boris is squat and strong with dark skin, maybe central Asian heritage. He's accused of stabbing two men to death. Frank asks him what happened but Boris won't talk about it. He'll only tell Frank that his father had both legs chopped off on a trainline when he was a kid, as if this is somehow a mitigating factor.

Yuri is skinny with an unhealthy pallor, but something in his eyes suggests he's a smart kid. He's in for a series of notorious robberies. The prosecutors say his signature weapon was the Taser, used mainly on conscript soldiers. Young men, gullible and new in town. And he went up to them – this is what the investigators claim – and patted them on the back then zapped them in the neck. He zapped them, they went crumpling to the ground like a ragdoll, then he rinsed them. The prosecutors say he targeted troops going back to their barracks, Tasered them on their doorsteps, then dragged them through the door and robbed their rooms.

Dima is in with Vitaly and Alexei. Vitaly is thirty-one and was an alcoholic on the outside. He lived with a woman in her fifties and existed on the fringes of society, without a passport or identity papers. They argued, he hit her. Because his arm was in a cast, he fractured her skull. He was arrested, she didn't press charges but because he had no ID card he was kept inside. He's been here five months and doesn't expect to get out anytime soon. Alexei, meanwhile, is in for armed robbery. He broke into the house of an associate – someone who owed him money – and beat the guy, then threatened him with a knife before scooping up a box of computer equipment.

Colin Russell's cellmate is a double murderer. He's a young guy, maybe twenty-one, sprung like a tight coil. He paces up and down the cell, stops, examines his muscles, does press-ups and sit-ups. He gets plastic bags and puts jugs of water in them, and lifts them in front of the mirror. Sometimes he punches the wall.

Colin – the 59-year-old Australian radio operator – asks the kid to sit on his bunk for a moment. The Russian stares quizzically at Colin then sits down. They try to talk. The guy doesn't speak much English but Colin manages to ask him why he's here. The guy says his best friend and his girlfriend were found in the front seat of his car, stabbed to death. But it was somebody else who did it.

Andrey Allakhverdov – the ship's chief press officer – has a TV in his cell, and every evening he watches coverage of his case on the state-controlled broadcast channels. It's a tsunami of shit being heaped on the heads of him and his friends. 'Do you see what they're saying about us?' he says to his cellmate. 'Can you believe this?' The news reports reiterate the claims made on NTV that the activists are agents for a foreign power, possibly employed by Western oil companies to sabotage Gazprom's drilling programme. And Andrey's cell-mate – who is charged under twelve clauses of the criminal code, including hooliganism – says, 'What do you expect? They're all state channels, just don't pay attention, it's okay.'

The Welshman Anthony Perrett is in with Sergei and Oleg. The prosecutors say Sergei mugged a stranger, ran away, got caught by a security guard, stabbed the guard and ran away again. He was married soon afterwards but two months later his wife left him, and now he's depressed. Oleg is from Ukraine. He was a chef on the outside, he makes beautiful salads, prepares them on a chopping board fashioned from an unfolded Tetra Pak and uses spices to season them with beautiful, rich flavour.

Anthony is thirty-two years old, a tree surgeon and director of a renewable energy company. Back home in Newport, he would tell people he was attacking climate change 'in the same way Wile E. Coyote tries to catch The Road Runner'. Before sailing for the Arctic he was working on developing a wood gasifier to run his forestry truck off a charcoal kiln, and a 3D-printed river turbine to generate remote electricity.

He's also a talented artist and loses hours sketching the view through the window. Oleg asks Anthony to draw something for him. He wants a giant bumblebee carrying a message. And Anthony says, 'Yeah, sure, okay.' He sits down and makes the sketch, and when Oleg sees it his face lights up. He adds a message, and that night he sends it to his girlfriend on the road.

'Who is she?' asks Anthony.

'My girlfriend? She's a hag, a crack whore, no teeth, but this does not matter because I will never meet her. I send her presents. She sends me little perfumed cigarettes.'

Anthony nods. And he's thinking, sure, I get that. Aesthetics are a luxury of freedom.

It's nearly 10 p.m. at SIZO-1, just before the lights go out, and Frank is sitting on the edge of his bunk, watching his cellmates Yuri and Boris construct the road.

Right now they're making the ropes. There are two different types of rope, but this one, the one they're making now, is a string made from the plastic bags that the prison bread comes in.

'Boris, what's that one called?'

The Russian looks up. 'This? We call this the *kontrolka*. This we need to make cells link together. Here, I show you.'

In one hand Boris is holding an empty paracetamol tube, and in the fingers of his other hand he's holding a broken razor. He slices off the end of the tube. Now it's a hollow plastic cylinder. He pulls a plastic bag through the tube, draws a pencil from his top pocket and ties the bag around it. He grips the tube in his hand, Yuri pulls the bag and Boris turns the pencil. He turns it and turns it so the bag twists. Yuri pulls the bag, shuffling backwards. It stretches and twists and stretches as Boris turns the pencil, using it as a spindle. Now the bag is a long frayed length of orange plastic, like trash on a beach, but twisting waves are running up the line as Boris turns the pencil, the plastic is thinning, it's getting darker in colour, getting denser and longer. It takes a few minutes, but forming before Frank's eyes is a strong deep orange string.

When they're done, the Russians start ripping strips from a bed sheet. They tie them together then attach the thin rope – the *kontrolka* – to the sheets.

'Boris, don't the guards punish you for ripping the sheets?'

'Our sheets get smaller. They don't care.'

'The big rope, what do you call it?'

Boris lifts the torn strip. 'This?'

'Yeah.'

'This is the *kon*.'

'*Kon?*'

'It means … what you say? Like a horse. It means … stallion.'

Frank nods and looks down at Yuri. The other Russian is on the floor of the cell, rolling up a sheet of newspaper. Every few days a paper is delivered to the cells but it's a state organ, absolutely pointless, no real news. Now Yuri is kneeling over a full page, rolling it tightly into a tube. He rolls it on the floor then stands up and rolls it on the wall. He rolls it and rolls it, taps the end and rolls it again until he has a stick about a metre long. Then he pushes a bent nail into the end.

He tears strips from another plastic bag, and he wraps those strips around the newspaper stick and melts the plastic with a match so it's sealed. It's as stiff as a truncheon now. Frank thinks you could do some damage with it. Then Yuri takes the thin rope – the string made out of plastic bags – and he attaches it to a bag with a bar of soap inside and hangs that bag off the bent nail.

Yuri hands the contraption to Boris, who walks over to the window. He slides the stick through the bars and leans forward as far as he can go. Frank jumps off his bed and stands behind him, peering over his shoulder.

Boris shouts out and a guy in the cell next door shouts back. That guy puts out his own stick. Frank can just about see the tip with a hook on the end hovering in the dull light.

Then Boris flicks his wrist and the bag of soap arches through the air, carrying a trail of string. The guy with the other stick tries to catch it with his hook but misses. Boris pulls in his stick, reattaches the bag and tries again. And on the fourth attempt the guy next door catches it and shouts, '*Doma doma!*'

He pulls the string through until he's got hold of the thicker rope – the torn sheet. Now their cells are connected.

For twenty minutes Boris does this in every direction, feeding ropes to the left, right, up and down. And everybody's doing the same across the wall, shouting, '*Doma doma!*' – 'It's home it's home!' – when they catch a string with the pole. When the ropes are in position they attach a sock to each line and soon the socks are going back and forth, up and down. An internet made of ropes.

Now the prisoners are banging on the floor and the ceiling. Frank worked out the thumping on the first night. Two bangs means, 'You've got mail.' *Bang bang bang* means, 'I have mail for you.'

To stay up until morning, Boris and Yuri brew a drink they call *chifir*, which they sip as long as the *doroga* is running. They take fifty grammes of tea, boil it up in a mug, take the tea out, add another fifty grammes of tea and boil it again. Then they add more fresh tea and boil it again – and again and again – until it's thick, like a soup. For some of the prisoners, *chifir* is not strong enough, they prefer to drink a concoction they call *kon* – again, it means 'stallion'. It's the same as *chifir* but with ten spoons of coffee powder added, and a splash of condensed milk.

Frank lies back on his bunk and watches the road come alive, counting the number of messages coming through his cell. Tonight is quiet, maybe a hundred notes. But on a busy night there are three or four hundred, and on those nights Boris and Yuri are absolutely buzzing because they have to drink themselves into a stupor with the *chifir*.

The road is against regulations. All pre-trial investigative detainment should mean total isolation. 'But the road is our revenge,' says Boris. 'All the things we do, the illegal things, they give us self-respect. We resist the rules. And there is solidarity in resistance.'

Roman is at *gulyat* – exercise hour – with his young cellmate. The kid shouts over the wall, explaining to somebody unseen that one of the Greenpeace prisoners is in with him. A commanding Russian voice comes back. 'Listen to me, and listen attentively. When you return to your cell, tell the guard this. Tell him I have instructed that you must break out. Be clear to use those exact words. Do not say anything more, do not ask him again, just wait. Do you understand?'

'I understand.'

When the *gulyat* is over, Roman and the kid return to their cell. The kid pulls a lever that drops a flag in the corridor. A guard opens the hatch.

'What do you want?'

'The *kotlovaya* has instructed that I must break out.'

The guard says nothing, the hatch slams closed. Then in the early evening the door swings open and the kid is told to pack his things and leave. He shakes Roman's hand, wishes him luck and disappears.

And that is that. An act of compassion by the *kotlovaya*, to protect Roman Dolgov from the taint of being a woollen guy.

*

On the second floor of Murmansk SIZO-1 the eight women from the *Sunrise* crew are being held alone. There is nobody to tell them about the road, they have no cellmates to explain this place. They don't know why the prisoners spend all night banging and thumping and screaming.

When Alex sees a rope dangling outside her cell with a bag hanging from the end, she jumps with surprise. She gets up from her bunk, wraps herself in her purple ski jacket and cautiously, silently, she approaches the window. The bag is small, the size of a fist, and it's swinging gently back and forth. She comes closer, stands up on her toes and looks into it. And she sees it's full of white powder.

Whoa! Cocaine. Okay, don't touch it. Do. Not. Touch. It. She backs away from the window. Okay, I've got these absolute nutters banging on the ceiling above me. They're obviously all fucking high on coke. And they're dealing. They're trying to sell me drugs, maybe half a kilogramme of cocaine. Be careful now. You've got to play this right.

She edges forward, bends down, tries to look up to see where the rope is coming from. And all the time the bag is silently swinging in a narrow arch in front of her window. Then suddenly the rope twitches, the bag is pulled up and the coke disappears.

Alex retreats to her bunk. The screams and shouts and bangs and crashes are exploding all around her, from above and below. She stares at the ceiling, her heart racing. If she'd had a cellmate she would have known that one of the Russian prisoners on the third floor was offering her some sugar for her tea.

Solitary confinement makes the hours feel like days and the days feel like weeks. The women maintain their sanity by constantly tapping to each other, using the code they agreed on that first *gulyat*. Camila, Sini, Faiza and Alex tap for hours. Each conversation takes an age to conduct, with a single sentence taking five or ten minutes to tap out. In an era of instant communication these exchanges take on a poetic quality, where every word has great meaning.

They greet each other at 6 a.m. as the porridge comes, and again at 8 a.m. when the prison falls silent and they grab what sleep they can. Then from 11 a.m. they're tapping constantly.

> *good morning how are you all*
> *am seeing lawyer today*
> *is there news*
> *everybody talking about us*
> *my lawyer said its big on british german dutch tv*
> *my lawyer said they closed road outside Russian embassy*
> *in buenos aires because so many people protesting*
> *wow*
> *omg*

The women tap to each other all day. And when they're not tapping, they're dancing to the music channel on their TV sets. There's an evening show called *Bridge in Time* – a compilation of timeless tunes from the sixties onwards. The women throw themselves around their cells, drumming on the radiator pipe and thumping the walls to the Beatles, the Stones, Led Zeppelin and Michael Jackson. Anything to feel a connection with other human beings.

The highlight of the day is the *gulyat*. Every time it gives them a surge of energy to actually talk to each other. They have to shout over the wall, but that hour can be joyful. They pull themselves from their fear and depression by sharing any good news they've heard. They've been told that people across the world are standing up for them. Their lawyers and the consuls from the embassies have told them their fate is a global news story. And when they share all this they feel something electric in the air. It's pride, and it flows over the walls.

Even now, facing perhaps fifteen years in this place, they tell each other they don't regret the protest. They know their friends on the outside are fighting for them. They know they did the right thing. 'And if we did the right thing,' Camila shouts out, 'then what can go wrong?'

She's twenty-one years old, the oldest of six children, and apart from holidays in Uruguay and a trip to the USA, this is her first time abroad. Camila is classically attractive with olive skin and long brown hair – an archetype of Latin America. On her first day in this place the guards confiscated the silver ring she wore in her nose. She grew up watching National Geographic documentaries on Argentine television, spending hours in front of the screen staring at the images of animals, savannah and rainforest. For years before joining the *Sunrise* she would lie awake wondering when she'd see for herself the creatures and lands featured in those programmes.

A year ago she was working in a fashion outlet, selling clothes to rich women. She hated that. Later she worked as a receptionist at an English language institute. Then suddenly the call came in from Greenpeace. They needed climbers for a direct action protest in the Arctic. Camila told her boss she was leaving. 'Sorry, there's something I need to do. One of those once-in-a-lifetime things. I quit.'

TWELVE

Frank is standing in a small room with brown walls in the depths of Murmansk SIZO-1. In front of him is a Russian man dressed in full camouflage fatigues, a peaked military hat and heavy eighteen-hole combat boots. The man is wearing Reactolite glasses, the ones that turn into sunglasses when it's bright. At his side, swinging from a finger, is a thick black baton.

He is the prison's resident psychologist.

Frank bites his lip and eyes the stick then slowly, cautiously, he lowers himself into a chair opposite the shrink. The guy looks like he's about to be deployed to Afghanistan. Frank's thinking, please, my friend, this is not a good look for a psychologist. You need to do something to soften your image.

The man sits down, lays the truncheon on the table, shuffles some papers then looks up. Frank scratches his nose. The man nods.

'You happy?'

'Not really.'

'You scared?'

'Sometimes, yes.'

'You want to … to harm Frank?'

'Do myself in? No. Not yet.'

'You like food here?'

'It's okay,' says Frank. 'I'm still alive.'

'People in cell, they good?'

'They're fine. Fine.'

The man nods. He jots down some notes then points at Frank and says, 'You, two-two-seven.'

'I'm sorry?'

'Two-two-seven,' says the psychologist.

'That's a cell?'

'You, two-two-seven.'

'It's a non-smoking cell?' asks Frank. 'I said I wanted a non-smoking cell.'

'Two-two-seven. Smoking? Smoking?' And the psychologist sucks on an invisible cigarette then points at Frank.

Frank shakes his head. 'No, no. No cigarette. I want a non-smoking cell. Are you saying two-two-seven is non-smoking?'

'Two-two-seven.'

'Yes, my new cell. Non-smoking. Let's go and look at it, let's go and look at two-two-seven right away.'

'Go?'

'Yeah, let's go to the cell.'

So off they go, the psychologist leading Frank down the corridor, his baton swinging at his waist, his boots stomping into the floor, the sound echoing down the hallway. But then he stops outside Frank's cell, the one he's just left.

'No no,' says Frank. 'This is not two-two-seven.'

'Uh?'

'Not two-two-seven. Two-two-seven is downstairs.'

He points at Frank. 'You. Two-two-seven. You.' He lifts the baton and jabs it into Frank's stomach, prodding him backwards until he's inside the cell, then the psychologist pulls the door closed. Frank's nostrils fill with the smell of cigarette smoke and wet laundry. The sound of stomping boots fades to silence. He stands in the middle of his cell, scratching his head, confused.

A few hours later a swarm of officials bursts in – five guards, a guy in a suit, a translator. And the one in the suit says, 'You must take this.' It's a charge sheet, it says he's officially accused of piracy under Article 227 of the Criminal Code of Russia. And right then Frank realises what the psychologist was trying to tell him.

He feels exposed. Very, very exposed. The psychologist was telling him he needs to get mentally prepared because piracy is ten to fifteen years. Frank is sure the guy was pointing at him and him alone, that he's being singled out for Article 227. They found his laptop, he thinks, and they probably found the flash card with the protest plan on it. He left it on the ship. He downloaded everything from the laptop onto the stick and taped it under the table in his cabin. But it's the FSB, isn't it? The KGB. They've found it, of course they have. Oh Christ, why did he even keep that memory stick? Why didn't he just throw it through a porthole into the sea? Then he remembers why, and the realisation makes him punch the wall and stifle a sob. That memory stick. That stupid fucking memory stick. He's going to lose contact with his kids because of that damn memory stick, and the only reason he didn't throw it overboard was because he had all his receipts copied onto it and he thought his boss would bollock him if he came back from Russia and couldn't do his expenses.

Yeah, the FSB have found it. And they've probably gone on Google and worked out that he's occupied oil rigs across the world. And now they're going for the ringleaders. They're going to let everyone else go free, but him and Dima and Pete are going down for piracy. That's what's happening here. And now he's not going to see his kids for fifteen years. They won't know who he is. They'll think he put his job before them. He'll be an old man by the time he gets out, his kids will be strangers. And Nina, his partner, she isn't going to wait for him. Why would she?

Frank starts pacing up and down the length of his cell. He lies down, gets back up, starts pacing again then stops, gets a book out, starts reading it but doesn't take in the words. His boy will be twenty-eight, his girl will be in her thirties. He'll be a stranger to them. He slams the book shut. He can't concentrate, he jumps back down and starts pacing again, breathing hard, scratching his head and chewing his nails, tapping his foot, sitting down then standing, pacing and pacing and pacing and not finding anywhere in the cell that's a good place to be. He's close to the edge now. Getting close.

Yesterday he started making a deck of cards using a pen and paper and a razor blade he snapped from a shaver. He cut fifty-two squares and meticulously drew each card. He thought if he could play patience then he could get through the days here. And now he stops pacing and retrieves the cards from his shelf, lays them out and tries to focus. Boris looks over at him.

'What you doing? What is this?'

'Cards.'

The Russian jumps down from his bed. 'Cards? What do you mean cards? No, no, no. *Nyet, nyet. Kartser.* You go to *kartser.* Me too.'

Kartser. It's the cooler. The punishment cell. And Boris is saying if you get found with those then we're all shafted, we're all going to the cooler. So the cards get put away, ten o'clock comes along, the lights go down, the road starts cooking. Frank writes a note to Dima.

Fuck man I've been officially fucking charged with two-two-seven. Are you charged with two-two-seven? What is it? Is it Piracy?

Ten minutes later there are three thumps on the wall. Boris pulls in the sock and hands a note to Frank.

Yeah man we're all being charged. Everyone, all thirty. This is a good thing, this means things are starting to happen man!

Frank's kneeling on his bed when he reads Dima's note. It takes a moment for its full meaning to sink in. He's not being singled out. It's not just him. He scrunches up the note in his hand and falls forward on the mattress then pulls the sheet over his head. He's wasted so much energy fighting the fear, thinking it's just him, but now the panic has passed. A minute later he's gently snoring into the pillow.

A stranger is standing in the doorway of Roman Dolgov's cell holding a clipboard. A chubby little man with a bushy moustache. He's not in uniform, instead he's wearing a shiny blue acrylic tracksuit.

'You,' he says. 'Stand up.'

'I'm sorry, who are you?'

'Popov. I'm the governor. The new chief. Arrived today. I'm in charge. And you are … ' He glances down at his clipboard. 'Dolgov. One of the pirates, yes?'

Roman stands up. 'No. I'm not.'

The man appears surprised. He consults his clipboard again then looks up. 'Oh, I think you are.'

'Have you read the law? The law on piracy?'

The man scowls. 'Of course I have. You think they let someone run a place like this if they don't know the damn law?'

'I think you'll find that law is not applicable to us.'

The man's mouth screws up. His nostrils flare. 'I think you'll find I don't give a fuck what you think. You probably think you're going to be a handful for me. Well let me tell you, I've dealt with a lot worse. You lot are pussycats compared to my usual stock.'

'That platform was not a ship. It's attached to the seabed. Legally you can only commit piracy on a ship.'

The man nods over Roman's shoulder. 'But those bars are bars, and really that's all that counts.'

He smiles, and Roman sees the flash of a gold tooth.

'And the law is the law, no?'

'Look, arsehole, I really don't want to have you here for a year before you go to the labour camp, but it's not looking good for you. And as long as you're here, I'll be here too. Best if you get used to the hierarchy, eh?'

And with that he raps his clipboard with a pen and swings the door closed.

SIZO-1 has its own code of ethics. A prisoner never sits down on a cellmate's bunk. But the first thing Popov does when he bursts into Andrey's cell is to sit on his bed and bark, 'Why the hell did you bring these people to us?'

'I'm sorry, who are you?'

'I'm the governor of this place. I *run* this prison.'

'Right.'

'Why did you bring them here?'

'To whom are you referring?'

'The *foreigners*. You've got an American, a Brazilian, Argentines, Frenchies. Six British. *Six!* Why the hell bring them to Russia, eh?'

'It was actually the authorities who brought them here.'

Popov snorts and rolls his eyes. 'Well we don't need that sort in Russia. Damn foreigners. What use do we have for Americans and all that lot, telling us what to do?'

A few minutes later Alex's cell door opens. She stands up. A man walks in. He's wearing a blue tracksuit. He looks around. Then suddenly his expression freezes, his eyes narrow, he points at the waste bin and explodes with rage. He's screaming in Russian. Alex flinches. She looks down and sees some leftover bread she threw away. Her feet shuffle backwards but a moment later she pulls back her shoulders, takes a step forward and shouts, 'I don't speak Russian, okay? I don't understand what you're saying.'

The man sniffs. He jabs his finger at the waste bin and yells in her face. Alex feels her legs shaking. 'It's not a problem with my hearing,' she says. 'I just don't understand Russian.' The man lifts his chin and stands on his toes, trying to look down his nose at her, but he's not tall enough. Instead he's standing before her like a ballet dancer affecting the demeanour of a dying swan. He spins on the ball of his foot and flounces through the open door. A moment later it slams closed.

By the time the road is up and running, the arrival of the new governor is all anyone can talk about. The Russians say he's been transferred from a prison in the north Caucasus, where he presided over separatist rebels from Chechnya and Dagestan. Rumour has it he ran a strict regime but got so many death threats he had to be moved.

THIRTEEN

It was a spring afternoon in 2010, three and a half years before commandos seized the *Arctic Sunrise*. BP's *Deepwater Horizon* platform was gushing tens of thousands of barrels of oil into the Gulf of Mexico every day,[44] and a group of Greenpeace campaigners were meeting in a Turkish restaurant in north London to discuss their response to the unfolding disaster.

John Sauven, the executive director of the UK office, surprised his colleagues by saying the reaction shouldn't even focus on BP or the Gulf Coast. From the inside pocket of his jacket he pulled a page from the *Financial Times* and unfolded the sheet of pink paper on the table. He'd circled a story detailing the plans of a British company, Cairn Energy, to explore for oil off the coast of Greenland.

'Imagine if a *Deepwater Horizon* happened in the Arctic,' he said. 'What we're seeing in America would be nothing compared to that. Arctic oil, that's where the frontline is. That's where we need to be.'

He said the lack of sunlight and the near freezing sea temperatures off Greenland meant oil spilled into the Arctic wouldn't break down in the way it did in the warmer waters of the Gulf of Mexico.[45] When *Deepwater Horizon* suffered a blowout, the only way to plug the leak was to drill a relief well (a process that eventually took many weeks, by which time several million barrels of oil had been spilled[46]). But in the Arctic, when the winter ice returns, the sea is covered in a sheet of white that would prevent the drilling of a relief well, meaning a blowout could see oil leaching into a fragile marine ecosystem for months,[47] even years. That oil would then gather in a black toxic soup under the ice and be carried by the currents around the pole, to eventually be deposited in pristine waters many thousands of miles away.

Sauven folded up the sheet of newspaper and proposed a plan. He said he wanted to requisition one of the Greenpeace ships and sail it north to challenge Cairn's drilling programme. He wanted to use direct action to halt the company's operations for as long as possible.

Greenpeace had already been campaigning in the Arctic for fifteen years. Activists had confronted BP's Northstar drilling operations in Alaska; Greenpeace ships had documented climate change impacts off Greenland and Svalbard; and Mads Christensen's team in Scandinavia had campaigned against the industrial fishing fleets taking their destructive methods to the Arctic.

But it was *Deepwater Horizon* that provided the spark for a new wave of action. With the world's oil giants moving into the melting waters above the Arctic Circle, Sauven got his way and the Greenpeace ship *Esperanza* sailed north to confront the new Arctic oil rush.

That summer Greenpeace played a cat-and-mouse game with the Danish navy – still the governing power in Greenland. The activists outpaced the Danish special forces

RHIBs and occupied the underside of the Cairn platform, forcing a temporary shutdown.

The following year Greenpeace returned to Greenland, this time with Frank Hewetson leading the logistics operation. He hung a pod and two occupants from another Cairn oil platform, halting exploratory drilling operations for two days. After the Danish navy removed the pod, Frank led a team of eighteen activists who scaled the platform and presented a petition to the captain containing fifty thousand names calling for an end to Arctic oil drilling. Frank and the others were arrested and helicoptered to Greenland, where they were jailed for two weeks.

WikiLeaks had just published a quarter of a million cables from US embassies across the world. They showed how the scramble for resources in the Arctic was sparking military tensions in the region – with NATO sources worried about the potential for armed conflict between Russia and the West. The cables showed the extent to which Russia was manoeuvring to claim ownership over huge swathes of the Arctic. One senior Moscow source revealed that a famous 2007 submarine expedition by the explorer Artur Chilingarov to plant a Russian flag on the seabed beneath the North Pole was ordered by Vladimir Putin's United Russia party.

Another cable detailed the lengths to which the USA was going to carve out a strong position in Greenland, and the concerns Washington had over Chinese manoeuvring on the island. Tensions within NATO were also exposed, as Canadian leaders privately expressed disquiet in the cables over the Western alliance's mooted plans to project military force in the Arctic in the face of perceived Russian aggression. Canadian Prime Minister Stephen Harper was quoted by diplomats as saying that a NATO presence in the region would give non-Arctic members of the alliance too much influence in an area where 'they don't belong'.[48]

For an earlier generation the Arctic was the stage for Armageddon. Across the vast silence of the frozen north, two superpowers had ranged their armadas of nuclear warheads against each other's cities. The skies above the pole were the shortest and quickest route to bring about the end of human life on Earth. And now, for a new generation, the Arctic was again the stage on which their future would be decided. A military build-up was under way as the oil giants prepared to colonise one of the last unclaimed corners of the Earth.

Greenpeace launched a campaign to create a legally protected sanctuary in the uninhabited region around the pole – an area where oil drilling and industrial fishing would be banned. A scroll signed by three million people in support of the sanctuary was planted on the seabed four kilometres beneath the pole, next to Chilingarov's Russian flag. Attached to the scroll was a 'Flag for the Future' designed by a Malaysian child in a competition open to the youth of the world.

In summer 2012 Greenpeace activists confronted the *Prirazlomnaya* platform for the first time. It wasn't pumping oil yet, but it was exploring for it. Six activists, including Sini Saarela and Greenpeace global chief Kumi Naidoo, scaled the side of the rig. They spent hours hanging off the *Prirazlomnaya* before being forced down by the spray from a freezing water cannon.

That same month, Arctic sea-ice cover reached its lowest level in recorded history. In summer 1979, there had been seventeen thousand cubic kilometres of sea ice at the Arctic,[49] spread out over an area the size of Australia. A generation later, there were just four thousand cubic kilometres left.[50] Scientists can't pinpoint precisely when the summer sea ice will disappear completely, but there will very possibly be open water at the North Pole in the lifetimes of those activists who'd just scaled the *Prirazlomnaya*.[51]

But what happens in the Arctic doesn't stay in the Arctic. It's the planet's air conditioner. That white sheet of sea ice reflects incoming solar radiation, keeping the globe cooler than it otherwise would be. As the ice melts, that sheet is replaced with dark water, which absorbs the sun's heat, warming the planet and melting more ice. It's like the difference between wearing a white or a black T-shirt on a baking day. So as temperatures rise, the ice melts, which causes more warming, which melts the ice. And as the ice melts the oil companies are moving north to drill for the oil that caused the melting in the first place.

Reasonable people have concluded that this is possibly insane.

Cairn Energy came up dry in the Arctic and pulled out.[52] Their operations were not helped by those two waves of Greenpeace direct action. But Gazprom was still trying to strike oil, and so was Shell.

The Anglo-Dutch energy giant – ranked the number two oil company in the world by size – sank $5 billion into its Alaska offshore programme.[53] But Shell was unable to explain how it would properly deal with an oil spill. The clean-up operation in the Gulf of Mexico required tens of thousands of people and 6,500 boats,[54] but when Shell published its plan for dealing with a blowout in the Arctic it featured a host of 'solutions' that were improbable, even farcical. Recognising that detecting pollution under the ice is almost impossible, the company proposed to deploy a dachshund called Tara, supposedly to sniff for spilled oil. The plan featured a photograph of Tara wearing an attractive red and green singlet and a GPS tracking collar.[55]

A senior official at a Canadian firm specialising in oil-spill response said, 'There is really no solution or method today that we're aware of that can actually recover [spilled] oil from the Arctic.'[56]

In July 2012 Shell's Arctic drill ship, the *Noble Discoverer*, ran aground on the Alaskan coast.[57] The company's oil spill containment system was so badly damaged in testing that a US government official revealed that it had been 'crushed like a beer can'.[58] When the *Noble Discoverer*'s engine caught fire the man in charge of Shell's Arctic operation, Pete Slaiby, told the BBC, 'If you ask me will there ever be spills, I imagine there will be spills.'[59] Then on 31 December 2012 Shell's Arctic oil platform, the *Kulluk,* hit heavy weather in the Gulf of Alaska and ran aground after attempting to transfer to a different port, partly to save money on a tax bill.[60]

Bowing to public pressure, Shell announced it was suspending its Arctic operation.[61] That left Gazprom. Putin's oil giant announced that it was determined to become the first company to pump oil from the icy waters of the Arctic.

The *Prirazlomnaya* was by now thirty years old. Its base had spent years rusting in a shipyard before being moved to Murmansk and capped with the scrap parts of a decommissioned North Sea rig.[62] Against Arctic Council guidelines, Gazprom refused to publish its plan to clean up any spill.[63] A short summary – posted on its website before being removed by the company[64] – revealed that Gazprom was ill-equipped to clean up a spill on the scale of *Deepwater Horizon*.[65] But that summer they were going to drill again anyway.

At Greenpeace a team was assembled to organise a return to the *Prirazlomnaya*. Sini volunteered to climb the platform again. Pete Willcox offered to skipper the expedition. Across the world sailors and activists were emailed and asked if they were prepared to join the *Sunrise* and sail for the Arctic on a mission to take on Gazprom.

Daniel Simons is being harassed by the secret police.

There's nearly always a car posted outside his legal headquarters in Murmansk. The two men in the front seats are wearing leather jackets and have Bill Gates haircuts. They take photographs of him and his colleagues, or observe them through the car window before scribbling in notebooks.

One night he's heading back to the hotel where he's staying. It's 1 a.m. He crosses the street and bends down at the car window.

'I'm going home now. Have a nice evening.'

The man at the steering wheel says nothing. Instead he pulls a pen from the inside pocket of his black jacket and writes something in his little book.

'Good night then,' says Simons.

The men stare straight ahead through the windscreen in silence.

Simons has been a Greenpeace lawyer for six years. He's thirty-three years old and lives in Amsterdam but speaks

Russian. After breaking off his holiday in Venice he rushed to Murmansk to recruit a legal team to defend the Arctic 30.

From the day after the *Sunrise* arrived in Murmansk, the Greenpeace team sees armed men in balaclavas wandering around town. Sometimes the men sit in the lobby of the hotel they're staying in. Sometimes they follow Simons and his friends in the street. It's the same guys who were standing guard outside the Investigative Committee that first night.

One night he's walking home when he hears the sound of crunching grit. He turns around. It's a taxi. A minute later it's still there, crawling along the road behind him. He approaches the taxi. The car drives away, but soon afterwards it passes him. Then a couple of minutes later it passes him again. It's driving in circles around him. His phone rings in his pocket, just for a second before the caller hangs up. Then it rings again. Two different Russian numbers that Simons doesn't recognise. It's past midnight. He has a brand-new SIM card. It's obvious what's happening here. The FSB is trying to link that SIM card to him, to check what his number is.

Two weeks have passed since commandos stormed the *Arctic Sunrise*, and by now one million people have written to Russian embassies around the world calling for the release of the crew. An Emergency Day of Solidarity sees 135 protests in forty-five countries across the globe.[42] There are demonstrations in the Russian cities of St Petersburg, Murmansk and Omsk. In Moscow there are pickets in front of the Kremlin and the FSB headquarters, and a protest in Gorky Park attended by the families of Roman, Andrey and the ship's 37-year-old Russian doctor, Katya Zaspa. Hundreds of people gather at the main harbour in Hong Kong to form a human banner that reads 'Free the Arctic 30'. In South Africa protesters come together at former apartheid detention centres. In Madrid supporters gather in Puerta del Sol with a replica of the *Arctic Sunrise*. In Senegal fishermen take to their boats to protest at sea in an act of solidarity. A year earlier they welcomed the *Sunrise* on its mission to preserve their fishing grounds from Western industrial trawlers. Now they're returning the favour.

Russian citizens have been officially complaining to the Investigative Committee and the General Prosecutor's office about the detention of the activists. They have to provide their personal information to make a complaint – their home address and phone number – and that puts them at risk of retaliation from the authorities. Nevertheless, thousands are registering their support for the *Sunrise* crew.

There are four global hubs from where the campaign is being run – Moscow, London, Copenhagen and Amsterdam. In London the Arctic 30 team has set up shop in a basement space at the Greenpeace office, and soon enough it resembles a cross between a teenager's bedroom and a military command HQ. From 7 a.m. until close to midnight the team sits hunched over computer screens, stuffing various iterations of fructose sugar into their mouths. On the wall there are three cheap clocks above childish hand-drawn flags – British, Dutch and Russian.

The campaigners' collective mental state is one of permanent kinetic stress, like that moment when the car in front brakes suddenly and a surge of fear floods the very centre of your brain. It's a constant communal condition, and soon enough their bunker is dubbed the Room of Doom.

The London hub is connected via a permanent video link to Mads Christensen's office in Copenhagen. Opposite him sits his wife Nora, the leader of the Arctic campaign before the arrests. She is now responsible for overseeing staff across the world working for the crew's release. The two of them make a formidable team. By 7 a.m. when London switches on the video link, the Christensens are at their desks discussing strategy for the day, having already got their two kids fed, dressed and into school.

At 10 a.m. every day, Copenhagen time, there is a core-team meeting. Faces from across the world appear in boxes on the screen: political operatives Neil Hamilton and Ruth Davis updating everyone on their dealings with governments; legal chief Jasper Teulings in Amsterdam giving the latest on his efforts to persuade the Dutch government to bring a case before the International Tribunal for the Law of the Sea; Rachel Murray, in charge of family liaison, relaying conversations she's having with relatives of the thirty; Ben Ayliffe discussing plans for demonstrations, vigils and direct actions; James Turner and Iris Andrews in Los Angeles talking about the involvement of filmmakers and celebrities in the campaign; Fabien Rondal with the latest from Murmansk; and the head of the media operation giving details of the massive publicity campaign being orchestrated across the globe.

This team is now manoeuvring a global campaign involving several hundred people working full-time across forty countries. Ensuring they all work to the same political, legal and communications strategies requires a huge effort of co-ordination that is unprecedented in Greenpeace history. The team leaders find themselves on eighteen different Skype chats – online discussion groups for press officers, legal strategists, family liaison, and every other subset of the campaign. The groups are indispensable tools for sharing information and enforcing discipline amid the fear that somebody somewhere might say something that crashes the legal strategy or offends a key national government. All campaigns involve risk, but with the freedom of their friends at stake, the sense of personal responsibility is sometimes overwhelming.

'We need to make them famous,' Mads Christensen tells them. 'We have to make every politician, every journalist, every business leader with investments in Russia, and every man and woman on the street know all about the Arctic 30. But we can't attack Putin. Not personally. We do that, they're fucked. We have to get them on TV and on the front pages, but not by hitting Putin. Instead we're going after Gazprom as a proxy for the Kremlin. If we can cause them enough pain we figure Putin won't think it's worth keeping our guys in jail. We need to give Putin a wide turning circle. We need to give him space to backtrack and release them. If this becomes a battle of wills between Putin and the West, we may never get them out.'

At Switzerland's St. Jakob-Park soccer stadium, FC Basel are about to face Germany's FC Schalke 04 in a UEFA Champions League group stage tie. Every seat in the arena has been sold, the TV cameras are in position, ready to capture the action for highlights shows that will be watched later that evening by tens of millions of people across the globe. In the sponsors' VIP boxes, suited executives are sipping on white wine and picking avocado canapés from silver trays, waiting for the game to start.

The evening's main sponsor is Gazprom. Putin's oil giant has paid to have its logo emblazoned across pitch-side hoardings at every one of that season's games, right across Europe. The television coverage is saturated with Gazprom advertisements; the Schalke players are on the pitch warming up in shirts bearing the Gazprom logo. It is a forty-million-euros-a-year [43] effort to detoxify the brand of Russia's state-owned oil company. And Andreas Schmidt is not happy about it.

'I used to practise climbing with Kruso,' he says, 'and now he was in the Arctic sitting in prison because of Gazprom. We wanted to show the public that what was going on was not right, that Gazprom is doing dirty business up in the Arctic while trying to polish their image in Europe by being a big sponsor. I know Kruso, he's a friend of mine. I was there because of him.'

Now Andreas is on the roof of the stadium. He and his team rig ropes then roll out a forty-metre banner. 'We had a little delay – we wanted to start just before the kick-off but we started just after.' They abseil off the roof, bringing the banner with them, unfurling it as they descend. The players are distracted, the attention of the crowd shifts to the sky, TV cameras spin away from the game. And with a final tug by Andreas, the huge banner catches the wind like a sail and fully unfurls.

GAZPROM, DON'T FOUL THE
ARCTIC – FREE THE ARCTIC 30

The referee looks up. It takes him a moment to understand what's happening, then he blows his whistle and calls the players off the pitch. Andreas and his team decided that afternoon that if the game was interrupted they would immediately end their protest, so they climb up the ropes and pull in the banner. A few minutes later the game resumes, but not before the cameras have caught images that will soon be broadcast around the world. Including in Moscow.

The next morning activists shut down every Gazprom station in Germany, locking themselves to the pumps.

This campaign can't go after Putin, but his oil company is fair game.

FOURTEEN

Dima is staring at the locked door of his cell, thinking, okay, there's this door, it's solid steel, twenty centimetres thick, the key to open it is the size of a shoe and I don't have it. Now, I don't want it to be closed. I want to get out of here. Sure. But if I keep banging my head against that door, that door is not going to open. But I will have a bloody head. So I'll still have a closed door and a bloody head, as opposed to having a closed door and no blood. Okay, so that means it's better not to bang my head against the door. And it's the same with the situation I'm in, the piracy charge, the fifteen years, the fear, the panic. It doesn't help me. And it doesn't help to beg for freedom. It changes nothing, so I'm just going to let it go. I'm going to get my head down and do my time, in the knowledge that people on the outside are doing all they can to get me out of here and there's absolutely nothing I can do to help them.

Ne Ver' Ne Boysya Ne Prosi.

Don't Trust Don't Fear Don't Beg.

For some of the activists, their cellmates are invaluable tutors in the techniques vital to psychologically survive the ordeal of incarceration. The Russians sit them down on their bunks and explain how to avoid antagonising the guards, how to stay on the right side of the bosses in the *kotlovaya* cells, how to communicate with their friends, how to fill the days and the long nights, how to hold on to their sanity.

Frank is sitting on his bunk with his head in his hands. His thoughts have been going round and round, faster and faster, and sometimes there's no way to stop them. He's thinking about his kids back home. If it's fifteen years he may be a grandfather before he gets out. His girl is sixteen, his son is thirteen. He could even die in here, then he'll never see his kids again. That could happen. That could actually happen. And if that happens …

'Frank, no. *Turma* racing. Bad bad.'

He looks up. It's Yuri, the quieter of his two cellmates. Because he's younger than Boris, Yuri is deferential to him. He rarely starts conversations but now he's looking at Frank and speaking softly.

'*Turma* racing. Bad, Frank. Bad.'

Frank shakes his head. 'What?'

'*Turma* racing.'

'What's *turma* racing?'

'This. Prison. This is *turma*. Russian word for prison. Racing. Your head. Round and round. Bad, Frank. Bad. Must stop. Not good.'

And Frank nods. Yes, Yuri's right. This is one of those moments when you're lying there and the vortex of panic is starting to spin, sucking you in, pulling you down to a dark place. You thought this thing ten seconds ago and now you're thinking it again and it feels even more frightening.

Turma racing.

In a cell down the corridor Dima can feel a tight fist of fear in his stomach. It's been there since that first interrogation at the Investigative Committee, and in his darkest moments he can feel it clenching tight. Sometimes it gets too hard to bear, when he's been thinking too much about that locked door that won't be opening anytime soon, or when he's been looking at the sky through the bars, thinking, will I ever see the sky without those bars? Will I ever see a sky that's not in squares?

In those moments he goes *uyti v tryapki*. It means 'into the rags'. When the prisoners want to turn off the external world, when they need to turn away from their lives, when they want to turn their backs on everything, then they smother their bodies with all their loose clothes, their towel, everything they have. And under that pile of their earthly possessions they face the wall on their bunk and go *uyti v tryapk*. That's what they call it, and Dima goes there often.

Joy and depression flood the cells in turn, but their arrival can rarely be predicted. When Dima finds out there's a well-stocked library here, he's ecstatic, this is great, he can be here for years, he'll read books during the day and at night he'll be on the road. Fuck this, man. I can do this! Then he turns on the TV and sees Medvedev, the Prime Minister of Russia, and he's saying, 'Well, pirates or not, these are very serious criminals. They're threatening the very livelihood of Russia.' And suddenly Dima is in freefall, it really is going to be fifteen years, and for the next hour he's *turma* racing.

For Denis Sinyakov, the writings of Aleksandr Solzhenitsyn serve as a vital crutch, giving him strength behind bars. The woman in charge of the library has read Solzhenitsyn herself and brings the great man's books to Denis's cell. His cellmate turns one of them over in his hands, perplexed.

'You're reading books about prison in prison?'

'Where better to read them?'

Denis saw Solzhenitsyn many times when he was alive, he photographed him, he covered his funeral. For Denis it's fascinating to read how he survived the gulag, and now Denis is comparing the conditions and the rules across the decades. And he sees that nothing much has changed.

Roman's first cellmate told him, 'At first you will count every minute here. Later you will count every hour. In three or four weeks you'll be counting the days. Then you'll count the weeks.' And it's true. Roman made a calendar and in the beginning he crossed out the days like Robinson Crusoe. At first he waited until the end of each day, and made a great ceremony of crossing it out. But now he finds he forgets.

Phil is in the *gulyat* box, staring at the sky. He's had a bad day, *turma* racing, and he scratches the words *fuck them all* on the wall. Afterwards he regrets it, he knows he needs to hang on to who he is. The next day he's back and sees one of his friends has rubbed out the first word and written the word *love* instead. And Phil thinks, yeah, that's the right attitude. That's how to survive this place.

The Greenpeace women, held alone on the second floor, have only their spoons, that pipe and each other. They're telling themselves it can't be fifteen years. Surely not. But then they see how they're being portrayed on TV, and their minds race towards the edge. They take up their spoons and tap to each other, working out how old they'll be when they're released if they get the full fifteen years. Alex will be forty-two. She taps on the pipe.

shit that means I can't have children

Camila taps back, her message reverberating along the pipe.

i'll be 36

Alex taps out a reply.

maybe we'll have to have sex with a guard
really?
i'm joking cami
well you do get two hours outside a day if you're pregnant
but they're all quite ugly
one of them is okay
which one?
the one who came to my cell today
oh him
yes
really?
no Alex, of course not. i'm not having sex with a guard to get pregnant
okay me too
good
good

FIFTEEN

Frank's lawyer reaches into his briefcase and pulls out a letter. 'It is from your wife,' he says.

'From Nina?'

'Yes.'

Frank's heart jumps. He snatches the envelope and turns it in his hands. It's bulging, full of sheets of paper, and Frank Hewetson thinks it may be the most precious thing he's ever held.

It's a Thursday morning and he and his lawyer are sat across the table from a senior officer from the Investigative Committee – a man memorable both for a streak of petty authoritarianism, and for an unnaturally enormous forehead that is capped with a surf of receding jet-black dyed hair. Most of the investigators wear cheap acrylic suits and this individual is no different, swathed as he is in a brown affair that creases violently when he makes even the slightest movement. A period of accusatory gesticulating at Frank leaves him looking like a crumpled wreck.

The man has just spent an hour telling Frank that the FSB now has proof of piracy, that many of the other activists have signed statements fingering Frank for responsibility, and that his only hope of avoiding many years in jail now lies with his revealing exactly who did what on the protest. Frank isn't sure he should believe the man but he won't incriminate his friends. Eventually the interrogation ends, and now Frank's lawyer is handing him a letter from his wife.

He never thought he'd feel such joy at being given a simple letter, but this is how it is since being locked up. He hasn't been sleeping well, the road runs all night and Boris and Yuri are loaded on *chifir* until 6 a.m. every morning, pulling the ropes, banging on the wall, screaming through the window. Already Frank is savouring the moment he'll lie on his bunk and run his thumb under the seal and pull out Nina's note. He'll read the letter slowly, savouring it, stretching out the time it takes to make it to the end. But just as he's turning the envelope in his hands, the officer plucks it from his fingers and slips it into the inside pocket of his scratchy brown suit jacket.

'No, you cannot have this. It must pass through our censors first. And you have not answered my questions.'

'What?'

'You can read letter when you answer my questions about criminal invasion of oil platform.'

'Just … come on, man. Give me back the damn letter.' He rubs a hand over the fuzz on his scalp. 'It's from Nina. I miss her.'

The investigator folds his arms, the suit bristles with static and multiple crease lines break out on its surface. He cocks his head and his eyebrows lift into the lower slopes of his forehead.

'No.'

'It's from my wife. Why can't I have it?'

'You are answering questions, not asking them. I am asking the questions.'

'You have to be kidding me.'

The officer sniffs, swings one leg over the other and narrows his eyes. 'When you tell me who was in charge of the criminal gang which attacked the platform, you may have the letter.'

Frank stares at him, at the thin mouth now rising at the corners as the man's face takes on an expression of supreme self-satisfaction, at the sweep of hair that only starts somewhere near the crown, at the saggy neck skin hanging over the collar of his shirt. Frank's lawyer is sitting next to him, two armed guards are standing behind the officer. And Frank thinks, shit, I'm fucked, I'm going down for fifteen years, it's happening, there's no way out, this is it.

He leans forward, eyeing the cop, biting his lip and making angry breathing noises through his nose, fulminating, trying to stop his mind. He's *turma* racing. He's close to the edge, the vortex is opening up. He's sucking in huge lungfuls of air through his flared nostrils, his knuckles are turning white as his hands grip tighter on the edge of the table. Then suddenly he hears a voice saying, 'Frank, are you okay?'

Frank looks around. 'What?'

His lawyer says, 'Are you okay? You've gone white.'

'No. No, I'm not okay. I'm fucking angry.'

'You need doctor?'

'Yeah, I need a fucking doctor.'

'Really? You need doctor?'

The guards edge closer, the investigator's smile collapses into an expression of panic, he's looking nervous, edging back from the desk. The cop clasps the top of his head with his hands and cries out in Russian. Frank doesn't understand him but it sounds like an expletive. The officer jumps to his feet and throws open the window, then he starts manically fanning the air in front of Frank's face with a copy of the criminal report into the boarding of the oil platform.

Frank rolls his eyes and makes a heavy gurgling sound in his throat as one of his legs goes into a spasm. The cop drops the report and pulls a lever arch file from a shelf. He opens it and uses it to fan Frank, and the look in his eyes betrays his fear that one of the Arctic 30 could expire on his watch. He drops the file and lifts a telephone receiver. Orders are barked, more windows are opened, Frank's chest heaves as he pulls a series of rasping laboured breaths. The door flies open and suddenly a doctor is standing in the middle of the room, his head turning from person to person as he searches for the patient. He rushes forward, applies a hand to Frank's head, sticks his ear against his chest then looks at the cop and shouts, '*Skoraya pomosh!*'

Ambulance.

Frank is carried outside and loaded into the back of a Russian ambulance, one of the guards jumps in next to him, the siren blares and they accelerate through the gates of the Investigative Committee headquarters. Frank's mind isn't racing any more. Now he's just confused. What's going on? Where are they taking me? Then he thinks, well, at least I'm getting a trip outside.

Five minutes later the ambulance skids to a stop outside Murmansk hospital, the door flies open and Frank is pushed into a wheelchair. The guard grabs the handles and bends down.

'We go to see doctor.'

'Yes, well, it is a hospital so I assumed that was next.'

'But you no escape. Understand?'

'I know I know, a move to the left or a move to the right is considered an attempt to escape and—'

'I shoot.'

'Yup, I got it.'

'Okay, good.'

'Yeah, don't shoot me, please.'

'A move to the left … '

Frank twists his head back to look at him. 'Yeah yeah, I know.'

'Okay.'

'Yup.'

He lifts the chair back, Frank grips the handles, the guard says, 'Okay, let's go!' then the wheelchair surges forward and bursts through the front doors of the hospital.

'You from London?'

'Yeah.'

They shoot across the foyer and take a corner at speed, two of the wheels lifting off the ground for a moment.

'Really?'

'Yeah, I'm from London.'

'Depeche Mode!'

'I'm sorry?'

'Yah yah, Depeche Mode. Depeche Mode number one. "Just Can't Get Enough". "Black Celebration". Depeche Mode number one!'

Now they're careering down a corridor, the guard's boots are making a slapping sound on the tiles as he powers forward with the wheelchair, doctors and patients are jumping into doorways, they flash past in Frank's peripheral vision.

The guard bends down to Frank's ear. 'You like Depeche Mode too?'

'Er … Depeche Mode number one?'

'Ha ha ha! Number one! *When I'm with you baby, I go out of my head, and I just can't get enough, I just can't get enough.*'

They skid into a lift, up one floor, then along a corridor at breakneck speed before Frank is disgorged into the arms of a cardiovascular consultant. He's immediately examined, the consultant expresses concern over Frank's heart rate, Frank tries to explain that he's been brought here by an armed joyrider who's just threatened to shoot him. The consultant nods, he doesn't understand, he doesn't care, he takes blood, he orders Frank to strip and lie on a bench. Frank takes off his top and lies down, electrodes are applied to his chest, tests are conducted, the guard plays with the safety catch on his pistol, the doctor disappears then reappears with a sheet of results.

'Your body is good,' he says. 'Maybe problem in head.'

Instructions are issued to the guard, Frank is loaded back into the wheelchair with the electrodes still stuck to his chest. He's spun around and launched into the corridor then into a lift, up one level then out into the psychiatric wing. They hurtle towards the door at the end, swerving to avoid another wheelchair coming in the opposite direction, wires trailing from his chest, the guard crooning over his head.

'… *and I just can't get enough, I just can't get enough.*'

They brake outside the door, the guard knocks then pushes Frank inside. Murmansk hospital's chief psychiatrist holds up a hand. He's on the phone. He's middle-aged with luscious grey hair, an expensive suit, a blue tie with red and white dots, a matching handkerchief in his top pocket. He finishes the call, motions to the guard to wheel Frank right in then fixes him with a superior stare.

'What's the problem?'

'Well I'm not really sure. I was being interrogated by the FSB and … '

'FSB?'

'Yeah.'

'You are one of the pirates?'

'Well, no, we didn't actually do it.'

The doctor shrugs. 'Of course.'

'No, seriously. We didn't.'

'The human mind is capable of convincing us of many things, most of all the things we want to believe.'

'Why am I *here*?'

'You tell me.'

'I felt unwell.'

'Then I will give you something for it.'

'I was being questioned by the FSB and I had a bit of a … I suppose it was a panic attack. Can you give me something to make me feel better?'

'Where do you think you are? This is a hospital, that's what we do.'

'Well maybe, er … *do you guys have Valium?*'

'Of course.'

'Can I, maybe … ?'

'I will write you a prescription.'

The doctor scribbles on a pad, rips off a sheet of paper and hands it to the guard. 'You'll get two a day from the prison doctor. Hope it helps.' And with that the man drops back into the seat behind his desk and smiles with paternal assurance. The guard spins the wheelchair around, bursts though the door and accelerates down the corridor. And half an hour later Frank is back at SIZO-1, being led to his cell.

He says to the guard, 'See ya, mate! Depeche Mode number one! Thanks!'

The guard turns around. He looks confused, surprised that a prisoner has actually smiled at him and said goodbye and thank you. He grins at Frank and says, 'Good luck, good luck my friend. Good luck, my Depeche Mode friend.'

That night Frank gets his Valium. He takes one and saves the other, and from then on he takes one every night, to get through the road.

SIXTEEN

It's burning a hole in his heel.

He knows it would be shown by television stations across the world, it could keep him and his friends in the news, unforgotten. But the memory card with the footage of commandos raiding the *Sunrise* is still sitting in that little slit in the sole of his shoe. Every day Phil shuffles to the *gulyat* in his boots without laces, contemplating how the hell he's going to get the damn thing out of SIZO-1.

It feels like a spy thriller, having this thing in his shoe. A clichéd plot device from a Hollywood movie. But for Phil Ball – a father-of-three from Oxford – it's real. He's smuggled the footage into a Russian prison, and now he has to get it out of here. His lawyer has just told him the Dutch government is taking Russia to ITLOS – the International Tribunal for the Law of the Sea – and the hearing in Hamburg is coming up soon. And Daniel Simons and his team have lodged appeals with the local court against their detention.

Phil knows the lawyers need the contents of his boot. It's obvious that footage would help them, they don't have to tell him that. If the judges at the international court could see what happened on the *Sunrise*, it could really change things, people would understand what happened, they'd see the activists with their arms raised, the aggressive takeover of a ship by masked soldiers.

So now it's breathing down his neck. He knows he's got to get the footage out before the hearing at the international court. It's vital. This is gold dust. There are people who look like pirates on that film he shot, and it's not the *Sunrise* crew. But it's in his shoe. He's in jail and it's in his fucking shoe.

Frank starts a diary in a small green exercise book. On the cover and the inside pages he sticks pictures of his family. The prisoners aren't allowed glue, so he uses dried toothpaste instead.

4th October
Sometime b4 10am I got summoned downstairs (escorted by 5 guards) to see Pavel my lawyer, 2 investigators + interpreter. I got a bit pissed when some mention was made of possible green vegetation drugs found in Doctors bag/cabin. I insisted on writing down that this was a cheap attempt to damage image of GP and besides the ship is Dutch + in international waters. Mind you the head investigator did offer to supply me with printed results of Premiership football. Had stroll around the Pig Pen and talked to Capt Pete over the wall. Also involved in ciggie packet transfer across cage between inmates. Guards got angry.
[later …] Was just taken across the yard to the Director of prison's office. The Director went on about tapping and Morse code between prisoners (ours) which I dismissed out of hand, mentioning no one had used such code since passing maritime college 20 yrs ago. On reflection I think

he may have been referring to Postman Pat and his black
and white cat [the road] during the quiet hours. I'm sure
it's common knowledge.

For the first week Sini Saarela, the Finnish climber who scaled
the side of the oil rig, has lived almost entirely on bread. As a
vegan she can eat none of the meals she's served through the
hatch in her cell door, and the prison authorities are refusing
to give her food she can eat. The guards say it's her problem
if she doesn't want to eat meat or cheese. Sini's lawyer fights
them hard on it and eventually they relent. Now every meal-
time she gets three big cold boiled potatoes, one carrot and
one slice of beetroot.

At first she's grateful, but soon she has to start throwing
potatoes away. There are too many of them. She tries to eat
all the potatoes but they're huge. Nine big cold potatoes every
day. She gets through four, maybe five, but the rest pile up in
her waste bin.

One night Popov, the prison governor, bursts into her cell
to conduct an inspection. Immediately the governor spots the
potatoes in her bin. His face contorts and he starts shouting
at Sini, pointing at the waste bin and screaming in Russian.

'I'm sorry,' says Sini. 'I don't understand you.'

Popov comes closer so the tip of his nose is nearly touching
Sini's face. Specks of spittle are flying from his lips and spat-
tering Sini's cheeks as he screams at her. His face is red. Thick
green veins are standing out on his temples. He's standing
on his toes to compensate for his short stature. Then he falls
silent, screws up his mouth, looks her up and down and
storms out.

Sini knows what he was saying. It was about the food in the waste bin. He was so angry. She sits on the edge of her bunk, shaken.

Popov doesn't usually make the cell checks so she thinks she probably won't see him again for a while. But the next morning the door swings open and Popov is standing in front of her, and a moment later he's pointing at the potatoes in the bin and screaming. Sini has never seen anyone so angry. Popov is shaking. He comes close to her, maybe twenty centimetres from her face, shouting and pointing at the waste bin, his moustache twisting and jumping. He's furious. Sini's terrified. She can't understand what he's saying, but she thinks he's threatening to throw her in the punishment cell.

Popov exhausts his fury and leaves. A shaken Sini sits on her bed and contemplates what has just happened. She can't eat all the food, it's impossible, the guards are delivering industrial quantities of potatoes and bread and almost nothing else. She looks around the cell then jumps up. She gathers all of the bread and takes it to the window. It doesn't open but there's a little hole for ventilation that opens sideways, enough space for a bird to poke its beak through. Sometimes the pigeons come into the space between the window and the bars so Sini crumbles the bread and leaves it there and watches the pigeons eating it. Okay, good, she can get rid of the bread. Then she grabs the bin, reaches in, pulls out a potato and drops it into the gap behind the bars. But the cold potato, this huge white hard potato, this potato that was boiled a week ago and kept in a fridge, is left untouched by the pigeons of Murmansk.

Frank Hewetson's diary

5th October Saturday

Another big sleep. I managed to filter out the wall banging, pipe tapping and shouting thru window. I'm only in touch with [Dutch chief engineer] Mannes now. He is in 410 and I'm in 320. Previously 305. Had a brief chat with Anthony over the wall. They all heard about the hospital trip and were a bit concerned. He reckons he has lost 9 kilos. Roman says he's lost even more. I get the impression weekends are slower in jail.

While Sini is trying to smuggle food out of her cell, Fabien Rondal – the leader of the ground team in Murmansk – is trying to smuggle letters in.

The rules of SIZO-1 dictate that letters to the prisoners must be posted to the jail and read by a censor before being given to the activists. Letters and packages posted from abroad take weeks to arrive in prisoners' cells, and even then the contents have been searched, sliced, scribbled on and blacked out. Rondal makes it a priority to short-circuit the system and open a communication channel between SIZO-1 and the outside world.

It seems impossible, but if anybody can do it, it's the 38-year-old Belgian. He comes from a family of opera fanatics, his brother is a professional singer, but Rondal's only brush with music was as a roadie for rock bands. He might once have had boy-band good looks but now his face is bearded. He speaks with a French accent and is a veteran logistics co-ordinator with a reputation for organisation and imagination. Four years earlier he led the team that evaded security at a summit of world leaders in Brussels by hiring limousines and posing as presidents and prime ministers before launching a protest on the red carpet in front of a bank of TV cameras. Three weeks before the *Prirazlomnaya* action, under cover of darkness, he installed two remote-controlled banners on the winner's podium at the Shell-sponsored Belgian Formula 1 grand prix circuit. Days later, as Sebastian Vettel was presented with the victor's trophy in front of a television audience of tens of millions, Rondal sent a signal from a mobile phone that activated the banners, so the German national anthem was accompanied by the sight of yellow fabric rising from the podium brandishing the message 'SAVE THE ARCTIC'.

So it comes as little surprise to his colleagues when Fabien Rondal finds somebody who can circumvent the censor and carry letters into SIZO-1. Soon, messages from family and friends – and from strangers who have read about the Arctic 30 – are pouring into the prison under the noses of the guards. The activists then have to hope the guards don't spot that the letters they're seeing in their searches have not been through the system. If they do, the punishment cell awaits.

Rondal promises he'll never reveal the identity of his smuggler, so for the purposes of this story we'll call him Mr Babinski.

The Babinski channel runs both ways, allowing the thirty to get letters out to their families, bypassing the weeks-long process operated by the censor. As soon as Mr Babinski gives Rondal the letters from the prisoners, the Belgian scans them, destroys the hard copies (in case of an FSB raid) then emails the letters to relatives and friends, who find messages from their captive loved ones dropping into their inboxes.

For Sini, the letters landing in her cell give meaning to her incarceration. They show her she's not alone. She gets letters from people she's never met, people who tell her they wish they could come to Russia and take her place in prison for a day so she could be free. And Sini thinks, I wish all the activists who are in prison because they fought for a better world could have the same support we have.

But she's still accumulating potatoes.

Nine new ones every day. She eats four, sometimes five, but that means just as many are being added to her uneaten stash. Popov hasn't returned yet, but it's only a matter of time before he conducts the evening cell check. And when he does, he's going to go crazy. Sini has tried everything. The pigeons won't touch them, and she's tried flushing them down the toilet but she almost blocked the pipe, and she won't be risking that again. This guy Popov is going to be really angry if she breaks the toilet system. She'll be sent to the cooler or moved to a different cell, and she really wants to stay in this cell because Camila and Alex are on the same corridor. So she starts hiding the potatoes in plastic bags. She ties the tops of the bags and hides them under the bed, behind the toilet, behind the clothes on her shelf. Her cell is full of contraband potatoes.

Two floors above her, Dima pens a letter to his friends in Stockholm.

I can only speak for myself, since I am isolated from my comrades. But from my perspective, if our action and the follow-up we are living through will lead to the undermining of the Arctic oil companies, and to get people all around the world, and especially in Russia, to understand the reality and urgency of the crisis we are attempting to avert – well that's well worth a few weeks or months or (sigh) even years behind bars. I am hoping that I can count on you, my friends and colleagues, to continue the campaign while the 30 of us are 'enjoying' this forced vacation.

As well as getting letters to and from the thirty, Fabien Rondal's team is using legal channels to get supplies to the crew. Alex is delving through her second delivery, pulling out desiccated fruit and sliced-up cheese, when she comes across a curious object. She holds it up, examining it quizzically. She turns it in her hands. It's metal with a handle and a stainless steel bar in a tight curl, and a plug at the end. Is this a hair curler? She plugs it in and examines it from every angle. It's a fucking hair curler. She tosses it back into the bag. Really? I'm hungry, I'm thirsty, I'm in jail, there are things I need right now, lots of things, but curling tongs aren't even near the top of the list.

The next day at the *gulyat* everyone is buzzing from the latest delivery of supplies, shouting over the wall, announcing to the other women what they've been sent.

'I got nuts! Oh wow, I never knew nuts tasted so good. When I get out of here I'm going to eat more nuts.'

'I got an orange.'

'I got beans!' Sini cries. 'I'm so excited I got beans. I ate them all at once, they were great.'

'I got peanuts and fruit,' shouts Camila. 'And I got a T-shirt. It's got handprints on the front, it was made by my family, all my sisters and brothers. I've been wearing it all morning, it makes me feel like they're here with me. I'm going to wear it to court, when we have the appeal.'

And Alex says, 'Well, I'm annoyed. Out of all the things they could have sent me, I get curling tongs. I mean, how impractical is that?'

'Really?'

'Seriously?'

'Oh my God,' says Faiza. 'They sent you curling tongs?'

And Sini says, 'Alex, what do they look like?'

'What do you think they look like? They look like curling tongs. A metal bar curled up, a wire, a plug. Curling tongs.'

'Alex, I think that's a water boiler. You use it to make tea.'

Silence.

'Oh.'

Every day Phil's cell is searched, his clothes are searched, he can't hold the camera card in his boot for ever, he needs to get it out. He thinks about handing it to Mr Babinski, but he doesn't yet know if he can trust him. It's one thing to lose a letter, but there's only one copy of this film in existence, and it could help him and his friends get out of here. And maybe Babinski has to pass through a metal detector to get out of SIZO-1. Instead Phil writes a note to Fabien Rondal and drops it into the Babinski channel.

When Thomas the Tank Engine's friend Harold dropped by for tea on the 19th, I was there with a go-pro. If you want to see it, let me know how, <u>who</u>, etc. There is no backup. Really very important. Please don't ignore this sentence Fabien. Looks exciting micro SD.

Why the hell Phil chose to adopt a code dependent on a working knowledge of 1940s British children's literature, only he can explain. But nevertheless the Belgian Fabien Rondal sends an email to Mads Christensen in Copenhagen with a correct interpretation.

To me, it means that Phil was the one who filmed the boarding of the ship on Sept 19th. Now it seems to me that Phil is saying that he still has the micro sd card with him now (in the jail?) and that he asks us for advice about what he should do with it.

Plans for Operation Extraction are laid.

Phil procures a matchbox. He removes the tray and turns the matches out, then he gets another matchbox tray and cuts the sides off. Then he pushes that piece of card smoothly inside the first matchbox, so now it has a false bottom.

He starts carrying his matchbox with him wherever he goes in the prison, testing to see if it's taken from him. He has a matchbox on him at *gulyat* and at searches, so the guards get used to the fact that he always carries this matchbox. He's trying to cover every detail, he won't leave anything to chance. Then he drops a note to Fabien through Mr Babinski.

Get a box of Russian matches and have someone in court for my appeal.

SEVENTEEN

It's 7 a.m. and the Room of Doom is filling up. From somewhere above, the team hears the sound of scraping soles dragging along the metal-grilled floor. Then a groan. The Greenpeace UK chief, John Sauven, is gripping the bannister. He closes his eyes and swallows, then breathes deeply. He's a tall man with sharp blue eyes and grey stubble on his chin. He's led the London office for seven years. It was his idea, back in that Turkish restaurant three years ago, to requisition a Greenpeace ship and sail north to challenge the Arctic drillers. He shuffles down the stairs to the Room of Doom.

'John, are you okay?'

'You look ill.'

'Do you want some water?'

'I … I went out with a Russian guy last night,' says Sauven. His eyes are bloodshot, his skin is wet with sweat. 'It was a former Kremlin adviser. I was pretty sure he was speaking for the Russian government, he was obviously very well connected. He told me he had fifteen minutes, that was all, so I met him at a hotel next to Buckingham Palace. I was … I … I was going to have tea with him but I got him a Duvel beer, it was strong and he really liked it so he wanted another one. Then he just started on the Scotch, and … and by the end of the evening we must have had about twenty-eight whiskies ... God, I feel dreadful ... We left at midnight then went back to his house to keep drinking. I was trying to get information from him, who he knew, how influential he was, whether he had any contacts within the Russian government that would be useful for us.'

He pauses. The team examines him with sympathy.

'Christ, I feel sick ... He said there's this jockeying for power inside the Russian system. He said hardliners, people connected to Gazprom, connected to the military intelligence services, they want to see us punished. And ... and there are people who are ... you know, who are far more liberal, with connections to the West, who are concerned ... Christ, I feel absolutely dreadful. You know, people concerned with Russia's image in the world. They see the Arctic 30 as negative for Russia, that they can't win this. Then he told me how pissed off Putin was when he was blamed for Pussy Riot. He got burned, he felt he was personally blamed. On the Arctic 30 he doesn't want blood on his hands, he doesn't want people to think he's responsible, but if ... if he's blamed then he won't back down, it's just not his style. So when we want to go in hard we have to go after his economic interests. We're doing the right thing with Gazprom. But Jesus, those Russians can drink. I haven't been to bed yet, he kept me up all night, whisky after whisky. I think I'm still drunk.'

Sauven shakes his head then shuffles off towards the toilet, leaving in his wake the wafting scent of stale booze and a clutch of embarrassed subordinates staring awkwardly at their shoes. He's spent the past few weeks meeting anybody he thinks can help the campaign. He doesn't care who he has to speak to. He's met the ex-CEO of one of the world's three biggest oil companies, someone he's sparred with publicly for years, and he's met someone who knows Putin's banker. He goes to art dealers who know the wives of oligarchs, and arranges meetings with the heads of hedge funds with Russian investments, seeking advice on how the new Russia functions.

His efforts are part of a huge behind-the-scenes intelligence operation dedicated to understanding the Kremlin's thinking. The set-up is run by Sauven, Mads Christensen, Kumi Naidoo and the campaign's political strategist Neil Hamilton. As head of the global campaign to free the Arctic 30, Christensen is the repository of all the hints, rumours and tip-offs that are coming in from across the world, from meetings with diplomats, foreign ministers, business figures, Russian liberals and even Kremlin officials. Sometimes the best indication of how the campaign is going is a simple glance at the video link to Copenhagen in the early evening to see if Christensen has cracked open a Carlsberg at his desk.

Mads Christensen is the epitome of concentrated Scandinavian cool. When he speaks to his core team, his voice is determined but composed. He's one of those leaders whose demeanour is as important as the decisions he or she makes. Something about him inspires trust and loyalty. It's not just what he says, it's how he says it. He projects calm at all times. Even when the campaign is on the back foot, when they are mired in crisis, he'll be found leaning back in his chair chewing gum with his feet up on the table at the Copenhagen hub, hands behind his head with his eyes closed, pondering how to outmanoeuvre the Kremlin. He wears colourful tank tops that would look ridiculous on most men, but somehow not on him. He has an intuitive understanding of the global media, a gift for strategic legal analysis and no tolerance for bullshit. A rumour circulates that he's not actually a human being. Instead, it is said, he's a computer algorithm, and Mads is in fact M.A.D.S. – Massive Automated Decision System.

He's running multiple sources across the globe. When contacts tell him who they're talking to, he can often find their names in diagrams of the Kremlin power structure. But the situation is confusing. Some Russians are telling him not to worry, to just sit tight, that this will all be resolved in a few weeks. But then he has other sources telling him this is going to get nasty, high-level people saying it's going to be years in jail. Christensen is told Putin is furious with Greenpeace and wants to teach the organisation a lesson.

Mads Christensen doesn't know who to trust – the people saying it will be over soon, or the sources predicting years in jail. Everyone wants to be part of a global story and he thinks some of his sources are mixing solid information with their own opinions. And he suspects he's being fed false intelligence by the FSB.

'There's an information war raging for sure,' he tells a core-team meeting. 'Who to trust and who not to trust, that is the question. It's making this situation very difficult to analyse.'

For the campaign's senior staff, this feels like playing a game of geopolitical chess. On one side of the board is a ragged band of amateur diplomats. Facing them are Putin and his security chiefs. The UK political director Ruth Davis wonders who might teach Greenpeace the rules of this global game, and she emails Martin Sixsmith.

Sixsmith is a familiar face, for years he was the BBC's Moscow correspondent, the guy who stood in Red Square wearing a thick coat and clutching a microphone as the Soviet Union collapsed. Since returning to London he's written a series of well-received books about Russia, including *Putin's Oil* – an investigation into the fight to control Russia's energy industry.

Ruth Davis is forty-six years old with wavy brown hair. She likes to holiday in the Arctic to see the northern lights, but since the ship was seized she's lived in the Room of Doom with a telephone plugged to her ear as she works her contacts in European governments. She doesn't expect Sixsmith to reply to her email. He's a busy man, the author of the best-selling book *The Lost Child of Philomena Lee*. It's been turned into a movie starring Steve Coogan and Dame Judi Dench and the film has just opened in London. It will later be nominated for four Oscars, including Best Picture. So the political director is astounded when Sixsmith writes back saying he has half an hour for her.

'Energy is what Russia's all about,' says Sixsmith, sipping on a cup of Darjeeling tea in a hotel bar overlooking central London. 'Without energy, Russia is nothing. Without oil and gas their economy would be in a dreadful mess, and the regime can only survive as long as the economy is doing well. For the years after Putin became president, things were good – economically I mean – and he was incredibly popular. Then they had a recession and you had tens of thousands, hundreds of thousands, possibly millions of people out on the streets demonstrating against Putin. He knows his own personal fortunes are linked to the economy, and the economy is oil and gas. He's put all his eggs in one basket and if he drops that basket then he's in big trouble. So when you guys come along and start protesting against the environmental effects of his energy strategy, then he's going to be extremely jittery. That's why you guys are in so much trouble.'

Ruth Davis looks up from her notebook. 'How much trouble?'

Sixsmith bites a chunk from a shortbread biscuit, spilling crumbs down his shirt.

'Big trouble. You have to understand how important the Arctic has become to Putin. Up until now Russia's relied on Siberia and this endless supply of oil and gas, but that's been extensively exploited, they're sucking it dry, so like the rest of the world he's looking for new oil, and the Arctic is on his doorstep. Russia was very quick to lay a territorial claim, they said their continental shelf extends right up to the North Pole so the Arctic is theirs. Remember when they planted that flag on the seabed? That's what that was all about. Then your friends go up there and say the Arctic's not Russia's, that the Arctic belongs to everyone, and of course Putin says, "No way." And now your friends are in jail.'

Ruth Davis lays out the strategy, how Greenpeace is giving Putin a wide turning circle, a chance to back down. Is it sensible? Are they on the right track? Sixsmith thinks for a moment then nods.

'Russia is a sort of half-European and half-Asiatic country. And in terms of attitude they're Asiatic, so not losing face is a really big thing. And the more you back Putin into a corner the more he'll come out kicking and fighting. You have to give understanding to the dilemma he's in, you have to give him room to make concessions, the opportunity to back down without losing face.'

'And who's calling the shots here?' Davis asks. 'We assume it's Putin driving events. He's making the big calls about where the case goes, right?'

'The move to lock them up will certainly have been dictated by the Kremlin. And in the Kremlin's terms, it's actually the right reaction, because the regime is in trouble. There were demonstrations in the streets last year after Putin's latest inauguration, protests in St Petersburg and Moscow. And what do you do when your regime is in trouble? You look for someone to blame. Traditionally it's always been the Americans, it's been the CIA. I bet they're saying on TV that you guys are funded by the CIA.'

'Oh yes.'

'Yeah yeah, and they'll be saying the CIA is funding you in order to embarrass Russia, because the Americans will lose no opportunity to do down the great Russian people. It's as if the Cold War never ended. Now of course we think that's ridiculous, and I suspect you don't get funding from the CIA— '

'If I told you I'd have to—'

'—you'd have to kill me, sure. But in Russian terms it's the right strategy to adopt, because if you demonise somebody who's challenged your regime—'

'By the way Martin, we're not funded by the CIA. The CIA hates us too.'

'I know that. What was I saying? Yes … if you demonise your opponent and show you're tough with them, like they were tough with Pussy Riot and your guys, then it's clear that the Russian state is standing up for the Russian people against its external enemies. You, Pussy Riot, the CIA. It's a political PR strategy *and* an energy strategy. You need to understand that. Putin's PR rating rose tremendously around Pussy Riot, people completely approved of that. Okay, so the metropolitan elites in St Petersburg and Moscow might think it's a bit hard sending those poor Greenpeace protesters to jail, or sending those girls playing their guitars to jail. But in the provinces people love that. Ivan and Masha out in the sticks, they'll be loving this.'

'Really?'

'You bet.'

He looks at his watch. Time is up.

'You have to go?'

'I'm afraid so.'

Ruth Davis folds her notebook. 'Thank you so much.'

Sixsmith shrugs. 'No problem. And don't worry too much. He won't keep them in for the Olympics.'

'No?'

'Well, he might. But if he does, he'll regret it.'

The campaign is deliberately not talking about Russia's Winter Olympics, which are scheduled to start in four months. The diplomats are telling Greenpeace that the Sochi Games are Putin's baby, that he's so closely associated with them that threatening to push for a boycott or organising protests there would back him into a corner and he'd come out swinging. Phil already used Mr Babinski to smuggle out a design for a Sochi Arctic 30 campaign – the Olympic rings as handcuffs – and Ben Ayliffe, head of the Greenpeace global oil campaign, is working on a strategy document for how the campaign will use the Games if the activists are still inside when they start. But, like Putin-bashing, right now the Olympics are taboo in this campaign.

Ruth Davis walks to the Tube station, reading back through her notebook, satisfied that their political strategy is the right one but shaken by the confirmation of something she suspected – that they're not even players at the chessboard, instead they're *on* the board, pawns in a much bigger game. The way Sixsmith explained things, Arctic oil is integral to the long-term survival of the Putin regime, it's Putin's golden goose and Greenpeace just kicked it.

The next day Kumi Naidoo makes contact with an influential figure in the Russian opposition movement. Naidoo gives him the codename 'Bagel'. When a meeting is secured Naidoo's office tells him, 'You have to go to the airport to have a bagel,' and he rushes off to meet the man in an arrivals lounge. Bagel tells Naidoo that Greenpeace must appear strong. If the organisation pleads for the release of the thirty, they'll never get out. Bagel says he respects how the campaign has been run so far.

'Putin doesn't respect weakness,' he says. 'You must never use the language of begging.'

Nevertheless, the campaign team starts working on a draft statement, to go in Naidoo's name, apologising for the protest and undertaking not to return to the Russian Arctic. 'Just in case,' says Christensen. 'It's not something I ever want us to release, but if we have to, if that's what it takes to get them out, then let's at least have something ready.'

But for the campaign team there is simply no bearable way to write to Putin using the words 'sorry', or 'we apologise', or 'we regret'. And, what's more, if Greenpeace does apologise then some of the thirty will never forgive the organisation, even if it secures their freedom. Their families, on the other hand, will never forgive Greenpeace if they don't.

For days they play with the language, but it's impossible. Eventually Martin Sixsmith is consulted. Does he know any English phrases that sound contrite when translated into Russian but which in English are less cowering? Sixsmith laughs. 'You know, if it comes to you having to say sorry, you don't get to write the apology. Putin's people write it for you. You just sign it.'

Vladimir Putin grew up in a tiny apartment in Leningrad with a picture of Felix Dzerzhinsky – the founder of the Cheka, forerunner of the KGB – on his bedroom wall. He led a thuggish childhood, getting into fistfights and abusing teachers. 'I was a hooligan,' he later admitted – an irony not lost on the numerous opponents of his regime who have been accused of hooliganism, the Investigative Committee's catch-all charge for political dissent.

In 1968 – the year Pavel Litvinov lifted the Czechoslovak flag in Red Square – the sixteen-year-old Vladimir Putin applied to join the KGB. He later said he was inspired by the TV show *The Sword and the Shield*, in which secret policemen broke down doors the length and breadth of Russia to protect the nation from its enemies. The KGB told the boy he was too young to join,[66] so instead he went to university to study law and successfully reapplied seven years later. His career progression was pedestrian but he eventually achieved a foreign posting in Dresden in East Germany, from where he witnessed the collapse of communist rule in Eastern Europe in 1989. Before the revolution was over he'd burned so many secret files that he broke the KGB's incinerator.[67]

In 1991 he left the service, although he later said, 'A KGB officer never resigns. You can join but you can never leave.'[68] As Russia slipped into economic and political chaos in the 1990s, Putin worked for the mayor of St Petersburg, securing for himself a reputation as an effective high-level bureaucrat.[69] By 1997 he was a government official living in Moscow and was taking an advanced economics degree. His dissertation, which extended to 218 pages, cogitated over the role of natural resources such as oil and gas in the Russian economy. The text was later declared classified, but Putin also wrote an article for a mining journal in which he summarised his thoughts. He wrote that Russia's energy policy – indeed the return of the nation to great-power status – depended on the management of the country's natural resources, which he valued at $28 trillion. Putin wrote of the urgent need to create Russian oil and gas companies that could compete with Western corporations. The government didn't need to own them, he wrote, but it should be able to determine their long-term planning.[70]

In 1998 President Boris Yeltsin appointed Putin to head the FSB – the main successor organisation to the KGB – and a year later made him Prime Minister.[71] On 31 December 1999 Yeltsin resigned his office. He'd suffered illness for some time, been drunk at state occasions and epitomised the failures of the post-Soviet nineties. Addressing the nation he said, 'As I go into retirement, I have signed a decree entrusting the duties of President of Russia to Vladimir Vladimirovich Putin.'[72]

Putin published a blueprint for his rule – *Russia at the Turn of the Millennium*. In it he stated: 'Russia cannot become, say, the US or Britain, where liberal values have deep historic traditions. Our state and its institutions and structures have always played an exceptionally important role in the life of the country and its people. For Russians a strong state is not an anomaly to be got rid of. Quite the contrary, it is the source of order.'[73]

While Russian state communism was dead, authoritarianism was not. Putin instituted a system of state capitalism in which wealth and power were concentrated in the hands of billionaire businessmen who were handed control over the nation's industries. As long as they did not challenge the regime and allowed the government to benefit from the riches flowing into their coffers, they were left to their own devices. When one of those oligarchs, Mikhail Khodorkovsky, owner of Russia's biggest oil company and the country's richest man, intimated that he might challenge Putin for the presidency, he was jailed on trumped-up charges. His company was broken up and distributed among Putin's supporters.

By 2007 about half of the Russian government's revenue came from oil and gas – double the amount when Putin took power[74] – and the Kremlin saw exploitation of the Arctic as key to its hold on power into the future. That year Artur Chilingarov planted his Russian flag on the seabed beneath the North Pole and claimed the Arctic for Moscow, provoking angry responses from other Arctic states.[75]

'I don't give a damn what all these foreign politicians … are saying about this,' said Chilingarov. 'If someone doesn't like this, let them go down themselves and try to put something there. Russia must win. Russia has what it takes to win. The Arctic has always been Russian.'[76]

EIGHTEEN

Sini writes a letter to the people across the world who are calling for her freedom.

> *The early winter is here in Murmansk, it has snowed a couple of days already. I spend a lot of time looking out through the window. When the sun shines it makes me think of you all supporting us, it makes me happy and makes me smile. When it is snowing, I think about the Arctic, the sea ice, the beautiful nature up here, and it gives me strength, it gives this all meaning ... I think about climate change, and I don't regret it, not even for a second. I would do all this again. All this and much more ...*
> *Sini*

But her cell is beginning to smell.

In every corner, under the bed, on the shelf above her bunk, there are plastic bags containing uneaten boiled potatoes. When she comes back from the *gulyat* the stench hits her square in the face. It's obvious she's hoarding potatoes, any fool could tell. Popov hasn't been back yet, but she knows it won't be long. She's desperate now, certain she'll be sent to the *kartser* the moment the governor does the rounds and sniffs the air in her cell. And so it is that one morning, just before a scheduled meeting with her lawyer, Sini scurries around the room pulling potatoes from bags and filling her pockets with them.

A guard opens the door and leads her down the corridor to the meeting room. Sini shuffles into a seat opposite her lawyer, Larisa, a provincial woman with a loud mouth and a garish trademark green synthetic top embroidered with golden dragons. She has already demonstrated great courage in her constant efforts to get Sini the food she needs.

'How are you?' she says.

'Larisa, I need you to help me.'

'I'm trying Sini, we're all doing our best.'

'No, I mean with something special. I need you to do something right now.'

'What's wrong Sini?'

'The head of the prison, he's crazy. He hates me.'

The guard is standing by the door, staring into the middle distance.

'Why does he hate you?'

'Because … because I don't eat all of his potatoes.'

'He hates you for that?'

'He went crazy, I think he wants to put me in a punishment cell. I need you to help me.'

'How?'

Sini eyes the guard then slowly, silently, she draws a potato from her pocket and holds it under the table. 'Please, take this. Take it out with you when you leave.'

Larisa's forehead scrunches up. 'Take what?'

'I'm holding it now.'

Larisa narrows her eyes then feels for Sini's hand under the table. Her fingers explore the contents of Sini's palm, a confused expression breaks over her face then suddenly she jerks her hand away and pushes her chair back.

'Sini,' she whispers.

'What?'

'You want me to smuggle potatoes out of the prison?'

Sini nods urgently. 'Yes, exactly.'

'Sini, I can't. If they caught me I'd be in big trouble. We both would. I'd be thrown off the case. I'd lose my licence.'

'So you can't do it?'

'No, Sini. I can't.'

'Really?'

'Smuggle potatoes? No. I can't.'

Sini slides the potato back into her pocket. 'Okay,' she says, trying to smile but wiping a tear from her cheek. 'I understand.'

The legal team has lodged appeals against the continued detention of the Arctic 30. The activists know the hearings are imminent, but none of them holds out hope that they'll be freed. 'Appeals don't work,' says Vitaly, Dima's cellmate. 'There's no such thing as an appeal. If they decide they're going to keep you, they keep you. If they decide to let you go, they just let you go.'

Roman's appeal is the day before his birthday. It's three weeks since the ship was raided and he's hoping for a present in the form of justice but, as soon as he sees the face of the judge, he understands everything. The man's face is frozen. Roman's lawyer tells him that the judge has a pre-printed text to read out when the time comes to deliver the verdict. He just needs to put down the name of the defendant, because everything has already been decided.

Roman is not at court, instead he is in a cell in SIZO-1 watching the hearing through a video conference link. After the evidence has been presented, the judge declares that he will now retire to carefully consider the merits of this complicated case. He orders the courtroom to be vacated. Then something odd happens. Everybody leaves the court so only the judge is left, but they forget to switch Roman off. So he's sitting there looking up at the judge. The man is supposed to be deliberating, thinking about what decision to make. But instead he clambers up onto his dais and swings his legs like a child in a playground. Then he jumps down, crosses the empty courtroom and collapses into the prosecutor's chair. He spins himself around then leaps up and falls into the defence lawyer's chair, spins around again, then sits down on a bench in the public gallery and takes off his gown. He sits silently for a moment, fiddling with the collar of the gown, then he glances up at the screen above his head. Roman is looking down on him with vague, perplexed amusement. The judge brings a hand to his mouth and cries, '*Dermo!*' – 'Shit!' – then jumps up and presses a button. The screen goes blank.

At the start of Kruso's hearing the judge accidentally starts reading out the judgement instead of the indictment, before the evidence is even presented. The defence makes a challenge against the judge, alleging she's biased. The judge goes away to consider the challenge against herself, comes back, declares that after careful consideration she's concluded that she is in fact not biased, that she was merely reading out preparatory notes for the ruling, and that therefore her impartiality is not in question.

As Phil walks into the courtroom he slips a hand into his pocket and draws out a matchbox. The guards push him into the cage and lock the door. Phil looks around, anxious for sight of a familiar face. He spots one of the Greenpeace support crew – someone who must remain anonymous, so we'll call her Mona. She stands on her toes and cranes her neck so she can see Phil over the cluster of photographers surrounding the cage. Their eyes meet, Mona nods and Phil nods back. She slips a cigarette from a packet and sticks it between her lips then shakes a matchbox and sniffs. One of the photographers steps back and examines the back of his camera, scrolling through pictures, and Mona slips into the empty space and thrusts a hand between the bars. Phil clasps it and shakes it firmly, a guard breaks through the photographers and pushes Mona back, snatching the cigarette from her mouth and remonstrating in Russian. Mona shrugs.

'*Izvinite*,' she mumbles. 'Sorry.'

'*Ne kurit*.' – 'No smoking.'

Mona holds up an apologetic hand and walks out of the courtroom, whistling to herself and shaking Phil's matchbox.

Phil's appeal is rejected and he's taken from the court in an *avtozak*. Kieron is with him, he's just had his appeal rejected as well. For both of them the trip to court was pointless, apart from the matchbox. That camera card was sitting in Phil's boot for weeks and now, finally, he's got it out of jail. But he's worried he was spotted. If someone finds that matchbox on Mona, they'll both be in a whole lot of bother.

'Hey, Phil.'

He turns to Kieron. 'Yeah?'

'I think I'm gonna propose to Nancy. If I get a phone call, I mean. I think I'm gonna ask her to marry me.'

'You're gonna propose?'

'Yeah.'

'To your girlfriend?'

'Jesus, Phil. Of course to my girlfriend. Who else?'

Silence. Phil sniffs. He doesn't want to say it, but he thinks it's a bad idea. He thinks Kieron should wait until he gets out of jail. If you do it over the telephone then Nancy can't say no, he thinks. You can't say no to someone who's in jail, so even if it goes well and you hear the right thing back, then later you'll have to deal with the paranoia. *Turma* racing. You're going to start thinking she only said yes because she didn't dare say no. So then the paranoia's going to build in your head, going round and round, and even though you're engaged you'll start thinking she's going to find a quiet moment to say, 'You know what, I've had second thoughts.'

But Phil doesn't say any of this to Kieron. His friend looks so happy just thinking about asking Nancy to marry him. So Phil nods and says, 'Yeah nice one. I mean, absence makes the heart grow fonder.' And that's the end of the conversation.

That afternoon Mona hands the camera card to Fabien Rondal. He uploads the footage onto a file-hosting site then he calls Mads Christensen on what he hopes is a secure line and tells him where to find it. The Danish campaign chief downloads the file, Rondal wipes it at his end and destroys the memory card. The campaign now has the footage for the hearing before the international court. Operation Extraction is complete. Mona strikes a match and lights a cigarette.

Day after day the *Sunrise* crew are taken to the courthouse to be told they must stay in jail. For Denis the experience of standing before the world's media in a cage in a Russian courtroom is weighted with irony. Until recently it was him wielding the camera, taking some of the most widely published photographs at the big political prosecutions. He recognises a lot of the journalists, some of them wave at him. He's surprised to see a friend from *Time* magazine there, but it's just more proof that his case is huge abroad.

On the second Friday of the hearings, it's Alex's turn. She's wearing her purple ski jacket and glasses, and as the verdict is translated she brings her hand up to her face to cover her mouth as her eyes fill with tears. Her father Cliff is watching live via the Internet. 'And that was the worst moment for us, when she was denied bail and she broke down a bit. That upset me. We couldn't believe what was happening. It was shock. Emotional shock.'

At Sini's appeal she holds up a small white postcard, on which she has written the words, 'THANK YOU FOR ALL THE SUPPORT'. She signs it with a heart. That evening she sits on her bunk and writes a letter to Fabien Rondal's ground team.

SIZO-1, cell 217

The court hearing was just like a theatre play where everyone (except me) followed the already set and decided manuscript. The question that has been bothering me the whole time of our arrest and all the investigations is that no one has been asking why I did it.

Climate change is the biggest and at the same time the most denied threat the world as we know is facing. And the dirty oil companies are taking advantage of the effects of climate change that are already visible … The oil companies are let to do whatever they want in the Arctic, that is said to belong only to some of the nations, but that actually is all ours since our future depends on it … There is no time to wait for international climate negotiations that practically lead nowhere … The oil companies are going to Arctic now and they are threatening the future of the Arctic, the climate and coming generations now, as I sit in this bloody prison.

I go and fight for the Arctic because I see no other possibility in the current situation. And as a person being from the Arctic I see that that is where my responsibility for the climate battle is.

I do not regret what I did. And if we went back in time, knowing about what our action would lead into, I would still do the same again.

Climate change is the one that doesn't forgive. And Arctic is what we cannot get back if we lose it. Justice comes and goes, freedom is there always if you just decide so.

Sini.

The *Arctic Sunrise* near the *Prirazlomnaya* oil rig, on its way to protest against offshore drilling in the Arctic.

Dimitri 'Dima' Litvinov. Born in Russia, raised in internal exile in Siberia, educated in America, he lived with his family in Sweden before sailing to the Arctic.

The 61-year-old American captain Pete Willcox on the bridge of the *Arctic Sunrise*, four days before the protest.

A Russian coastguard officer pulls a gun at the *Prirazlomnaya* protest.

Frank Hewetson stands under the Russian Spetsnaz helicopter as the first commando rappels onto the deck of the *Arctic Sunrise*. This photograph was taken by Denis Sinyakov. His camera card was then confiscated and only returned by the FSB 15 months later.

Activists and commandos flood the helideck. Phil Ball (arm raised, holding a camera) films the raid. He later hid his camera card in the sole of his boot.

Dima is pushed to the ground outside the bridge door, minutes after soldiers started landing on the *Arctic Sunrise*.

British activist Frank Hewetson at Leninsky District Court in Murmansk, where he was told he would be jailed while the authorities investigated an allegation of piracy.

Sini Saarela from Finland at her appeal hearing in Murmansk.

A smuggled image of one of the cells at Murmansk SIZO-1, where the Arctic 30 were held.

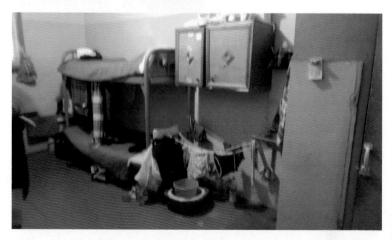

The outside wall of Murmansk SIZO-1. At night the windows would be connected by ropes that formed the *doroga* – the road.

Another view of the prison's exterior.

Mads Christensen with his wife and colleague Nora at the
Global Day of Solidarity in Copenhagen.

Two weeks after the ship was raided, 1,300 people marched past
the Russian Embassy in Helskinki, one of 135 protests held in
forty-five countries on the same day.

British activist Phil Ball at his bail hearing in St Petersburg. Sewed into his T-shirt are the words SAVE THE ARCTIC! in Russian.

Ben Stewart and Ben Ayliffe watching the live feed from court on the second day of the bail hearings.

Sini Saarela, Alex Harris and Camila Speziale.

Group shot of the Arctic 30. They are, from bottom left: Denis Sinyakov, Kieron Bryan, Roman Dolgov, Mannes Ubels, Frank Hewetson, Phil Ball, Ana Paula Maciel. From upper left: Iain Rogers, Sini Saarela, Camila Speziale, Gizem Akhan, Alex Harris, Cristian D'Alessandro, Hernan Orsi, Pete Willcox, Anne Mie Jensen, Faiza Oulahsen, Jon Beauchamp, David Haussmann, Marco 'Kruso' Weber, Ruslan Yakushev, Colin Russell, Paul Ruzycki, Alexandre 'Po' Paul, Dima Litvinov, Anthony Perrett. Missing are Francesco Pisanu, Andrey Allakhverdov, Tomasz Dziemianczuk and Katya Zaspa.

Pete Willcox with his wife Maggy. Before sailing for the Prirazlomnaya Pete sent Maggy a postcard, saying: 'If the Russians keep their sense of humour, I think this is going to be a fun action.'

NINETEEN

The appeals go on, the process grinds forward, always with a sense of impending, inevitable rejection. But the Kremlin does not only have the thirty in its sights.

Mads Christensen has sources telling him an FSB raid on the Greenpeace office in Moscow is a real possibility. Bank accounts will be shut down. Staff will be arrested and charged with complicity in the action. Any doubts about the veracity of the information fall away when the rental contract on the Russian office is cancelled with just three weeks' notice. It's a typical tactic familiar to Russian opposition groups. A decision is taken to start pulling out the foreign staff in Russia, even Daniel Simons, who's leading the legal response team in Murmansk.

In Moscow, on the balcony outside the Dance Hall, the head of Greenpeace Russia, Sergey Tsyplenkov, tells Laura Kenyon – a Canadian campaigner – what the sources are saying. The Investigative Committee is compiling a list of people it will potentially be investigating, and he assumes her name is on it. It's too risky for her to stay here. As a foreigner she's exposed. It's time to go.

Simons and Kenyon leave Russia. Everybody else waits for the Kremlin to make its move.

The following day nothing happens, but rumours of an assault on the Moscow office are swirling like confetti. Now sources inside the Russian government are warning that the crackdown is imminent, that the office will soon be shut down and staff arrested. Instead of bringing this saga to a close, it seems there is an appetite in the Kremlin to escalate.

Late in the afternoon Mads Christensen taps the microphone on the video link between London and Copenhagen. He asks Ben Ayliffe and the head of the media team to get on a secure line.

'Look,' he says, 'we've got a source, a really good one, someone who knows what's happening at the top level. I don't want to say who this is, and you don't need to know, but they're telling us it's going to happen. Maybe as soon as tomorrow. Office shut down, bank account closed, staff picked up by the FSB. We need to do something. Something that makes them stop and think. We need to make some kind of intervention so the PR hit they think they'll take – inside and outside Russia – makes the FSB think again.'

Ben Ayliffe and his colleague lock themselves in a room with a bag of pastries and a pot of coffee and thrash through ideas. Ayliffe leads the team organising demonstrations, vigils and petitions around the world. He's a twelve-year Greenpeace veteran with a passion for cricket, bird watching and shutting down polluting infrastructure using peaceful direct action. It should have been him on that ship but he hurt his back and Dima took his place.

Just over an hour later they have something for Christensen. It's a draft of a letter to Putin from Kumi Naidoo. Not much in itself, but this letter has a twist. It includes a serious offer by Naidoo to swap places with the Arctic 30.

> *Unlike the world leaders with whom you are more used to convening, I would not carry with me the power and influence of a government. Instead, I would come equipped only as the representative of millions of people around the world, many of them Russian, whose fervent wish is to see an early end to the continued imprisonment of the brave and peaceful men and women held in Murmansk.*
>
> *Were our friends to be released on bail, I offer myself as security against the promise that the Greenpeace International activists will answer for their peaceful protest according to the criminal code of Russia.*
>
> *I appreciate the risk that my coming to Russia entails. Last year I was part of a peaceful protest that was identical in almost every respect to the one carried out by my colleagues. In coming to Russia, I do not expect to share their fate, but it is a risk I am willing to take in order to find with you that common understanding.*

'But we need to send it tonight,' Ayliffe tells Mads Christensen. 'Moscow is four hours ahead and we need to hit the morning news there.'

Christensen rings off and reads through the letter, then he calls Kumi Naidoo.

'Mads.'

'Kumi, hi.'

'Hi.'

Silence.

'Mads, are you there?'

'Yeah. So, er … we have an idea.'

'Go on.'

'Well, we're trying to find something so morally powerful that the FSB can't shut us down, right?'

'Right.'

'And we think … we think maybe you could offer yourself up. In exchange for the others, I mean. It would sort of be sacrificing you on the altar of saving the Russian office. I mean, I know it's crazy, but what do you think?'

And Naidoo comes straight back. 'Sure.'

'Really?'

'Yup.'

'Are you absolutely sure about this? Because once you say yes, they could call our bluff. You know that?'

'It's fine, let's do it. I've actually been thinking the same thing for a while. I've been playing it over in my head for a week. I'm the boss, the buck stops with me. If I could swap with those guys in jail I would. Let's do it.'

Kumi Naidoo has been jailed before. When he was fifteen years old he joined the national student uprising against apartheid rule in South Africa. Kids across the country were walking out of school and taking to the streets to protest racist rule. Naidoo became a leader of the uprising, he was jailed, released, and forced to live underground. Eventually he had no choice but to leave the country. His offer to take the place of the thirty is a serious one.

The letter to Putin is delivered to the Russian ambassador in The Hague, the campaign sends out a press release, and the next morning it's a major story in the Russian media. Putin's spokesman says the President has read the letter but is power-less to intervene in Russia's independent judicial system. Around the world – but most importantly in Russia – it's known that Kumi Naidoo has made a personal offer to Putin to take the place of the Arctic 30.

The campaigners wait. Every time a new Skype message pops up from the Russian office it's quickly scanned as they look for news that the security services are raiding the Dance Hall. But nothing. No raid, no arrests. It takes another day for the Kremlin to react to the letter. But when they do, they play a card from the bottom of the deck.

'Okay, so the FSB found drugs on the *Arctic Sunrise*.'

'What?'

'Drugs. It's on the Investigative Committee website. Can you give us a comment? We're going live with it on the evening news.'

Tatiana Vasilieva, a 23-year-old press officer, lowers the phone and looks around. It's late, the Dance Hall is emptying, she was about to leave for home herself. The journalist on the end of the line is still speaking, she can hear his voice buzzing from the receiver. And then she hears another phone ringing, and another one. A moment later every phone in the room is ringing and her colleagues are reaching into pockets and bags for their mobiles.

'I'm sorry, say that again.'

'I said do you have a statement? We're going live in ten.'

'What kind of drugs?'

'Illegal drugs. That's what they're saying.'

Seconds later the BBC's Moscow correspondent tweets the news. The Skype groups explode with messages.

Aaron Gray-Block: Daniel Sandford @BBCDanielS
Russia's Investigative Committee now saying 'poppy
straw' and 'morphine' found on the @gp_sunrise

'Jesus Christ,' Mads Christensen mumbles to himself, staring at his screen. 'This is bad. This is very bad.' He unmutes the video link to London and taps the microphone. Eight heads look up. 'You lot seeing Skype? The Russians are saying they've found drugs on board the *Sunrise*. They're saying they found morphine and poppy straw.'

Faces duck below laptop screens then surface a moment later with wide, fearful eyes.

'Holy shit,' says Ben Ayliffe, shaking his head. 'That has to be bullshit.'

'It's a smear,' says Christensen. 'Total bullshit. Morphine and poppy straw.'

'I mean, poppy straw, that's crap. There's no way they found poppy straw on that ship.'

Silence, then Christensen says, 'What *is* poppy straw?'

'Er.'

'Ummm.'

Ayliffe types the words into Wikipedia, scans the page then looks up at the video screen.

'Oh man, it's opium. Raw opium stalks.'

'Who the hell sails on a ship with raw opium stalks?'

Christensen taps 'Greenpeace' and 'drugs' into Google News and sees the story is already getting pick-up. Western right-wing media outlets – many of whom have done nothing to cover the story of the Arctic 30 until now – are pouncing on the claim and posting their first dispatches since the arrests, with the prefix 'BREAKING NEWS'. He unmutes the microphone on the video link.

'Poppy straw anyone? What is it? Is it medicinal, or is it a recreational drug? We need to know as soon as possible.'

Laura Kenyon: In Russian this phrase that Google translates to 'poppy straw' is the same as what we usually mean in English if we say heroin. i.e. referring to the illegal kind of heroin.

'Oh shit,' says Christensen. The Room of Doom looks up. 'Okay, so Laura's saying the FSB are accusing the Arctic 30 of being on heroin.'

'Whoa!'

'Yup.'

They say a lie can go around the world before the truth has even got its boots on, and now the team is in a race against time to catch up with the FSB's smear and challenge it. The Western media hasn't used the 'H' word yet, but it's only a matter of time, and the implications for the thirty could be huge. Mads Christensen has a global campaign rolling here, but the FSB is trying to derail it with undiluted bullshit and the media is falling for it.

The campaign hits every Skype group out there, reaching hundreds of people, urgently demanding evidence that the Russian claim is a smear. Seconds later they're told that morphine is obligatory on all Dutch ships and that it's kept in the captain's safe. It would have been illegal for a Greenpeace ship *not* to carry morphine. But what about the opium? Within a minute they're being told that the ship was searched by Norwegian drug sniffer dogs before it sailed for the *Prirazlomnaya*, and a minute after that they're sent the certificate to prove it. Unless someone on the crew arranged a rendezvous with a poppy straw dealer in the middle of the Arctic Ocean, they can prove the drugs story is a lie.

Christensen's London-based media team bashes the information into a press release. In Amsterdam Daniel Simons watches the words appearing in real time on a Google doc and gives them legal sign-off as each sentence appears. They're desperate to get something out before the media accuses their friends of stashing heroin on the ship. Then twenty minutes after the FSB released its statement Greenpeace are sending out theirs, rebutting the smear in forensic detail and castigating the FSB for stooping so low.

Soon enough the media is running their corrective. Christensen types the words 'Greenpeace' and 'drugs' into Google News and reads reports saying the only drug found on the *Sunrise* was medical morphine. The tone of the story has changed. The FSB is being ridiculed. Russian Greenpeace campaigner Vladimir Chuprov – who'd be one of the first campaigners to be arrested in a raid – is quoted saying, 'Next they'll say they've found a pink zebra on our ship, or maybe an atomic bomb.'

Pavel Litvinov – Dima's father – is in no doubt about who's behind the smear. He's been expecting it. He's surprised it took them so long. 'I knew they would play with all these things, with drugs, that they would make it up even if they didn't find something. It was clear the command would come from Putin that this has to be done. Whatever they want, they will find. So I always had a fear they would say drugs.'

By the end of the day the story has died down. Mads Christensen comes on the video link. 'Well done everybody,' he says. 'I have to say that was extraordinary work. They tried to kill us today, but we stood up to them, we fought back and we survived. Today was a big day. Something important happened. This wasn't about drugs, this was about something even bigger. We're having a conversation with the FSB. This is what's happening, I think. We do something and they react to it, they do something and we react. This drugs thing is clearly a response to the Kumi letter. We sent it yesterday, then today they say they found heroin. Putin got the letter. That was his reply.'

Frank Hewetson's diary

9th October Wednesday

Just seen 20:00 news where the investigation team have claimed to have found 'narcotics' on Arctic Sunrise. Morphine of course. In the ship's hospital in fact. They are trying every trick to use the black arts of propaganda against us. If I wasn't banged up I'd be laughing.

A second day of global action is organised by the global campaign team. More than one hundred events are held in thirty-six countries involving nearly ten thousand people – everywhere from Mount Everest to Bangkok to Naples. The team in Murmansk plans a one-person vigil in the city centre, with a protester posing in a purpose-built cage made from cardboard and tape. The cage is stored in an enclosed yard at the rented building hired by the team as a headquarters.

Tatiana Vasilieva, the Moscow-based press officer, has travelled to Murmansk to help organise the protest. Under Russian law a demonstration involving more than one person requires a licence from the government[77] – a licence she'll never be granted. The Greenpeace plan is to put a lone protester inside the cage in front of the court building. The journalists attending the next appeal hearings will then see the person in the cage. But on the morning of the protest another press officer rushes into her room and sits on her bed, shaking.

'What happened?' asks Vasilieva. 'What's wrong?'

'The cage has been stolen.'

'What?'

'An hour ago when the team arrived to collect the cage it was gone. We spoke to the security guard of the building. He gave us the video footage from the CCTV cameras. We watched it and … '

'What? What does it show?'

'There were six men, all dressed in black and wearing masks. These men, you can see them, they're scaling the fence and moving in a straight line. Then they pick up the cage and carry it across the courtyard. They're either local freaks or the FSB. They have to be.'

Frank Hewetson's diary

11th October Friday

Got notice of my appeal being held on 15th Oct. Boris and Yuri are playing backgammon. I can't do that game. Just found out Boris has another 8–12 years to go but only 2 months left in this facility. Yuri says he has another 5–7. Puts things a wee bit into perspective.

12th October Saturday
I wonder what Nina [his partner] and the kids are doing this weekend. I really miss them. Nell [his daughter] is going through such a growth of maturity, ability and humour that I just can't help but feel I'm missing out on wonderful times. Every Saturday I miss is a Saturday I don't get to cycle down to Roundwood Park with Joe [his son] and Pluto [the dog]. It gets me deep down that these days are slowly slipping away from us. I love those two kids so very deeply. I'm so scared at how much they will seem to have grown + changed by the time I get home. These are low moments.

Since arriving at SIZO-1 Frank has examined every aspect of his life. He's raked over the decisions he took over many years, and reconstructed how he ended up on that ship. He wonders if he might have taken a different path. He remembers details of his childhood for the first time in decades. And he thinks about his father. Was he trying to live up to his dad's reputation? Is that why he joined the ship and sailed to the *Prirazlomnaya*? He often asks himself the question. He always envied his father because he had a cause, and environmental protection became Frank's cause. It gave him strength to know what he was doing was important.

Michael Hewetson would have been a formidable supporter of the campaign to free his son, had he still been alive. He was one of the legendary commandos dropped behind enemy lines the night before D-Day to secure Pegasus Bridge – the key strategic goal on which depended the success of the Allied invasion of Europe. The nineteen-year-old was in the thick of brutal battle for ten days before being wounded and shipped back to England, patched up and sent back to Europe, where he fought in the Battle of the Bulge and the crossing of the Rhine.

After the war he became a French teacher. He could never handle loud bangs and wouldn't have balloons in the house. It took Frank's mother years to get Michael to walk on the pavement instead of the middle of the road – a legacy of street fighting in Normandy.

Frank knew his father did something extraordinary in the war, but Michael refused to talk about it. In 1991 he returned to Pegasus, and his family came with him. They went to the cemetery. He stopped at a line of graves and broke down. Frank followed after him. They were all eighteen, nineteen years old. Names from Michael's past. But still he didn't open up.

It was on a family holiday in Spain that Michael finally told Frank about those ten days in Normandy. He said it was terrifying, it was chaotic, at times more brutal than the Russian front. Then he told his son a story he'd never told anyone before. It was day four after D-Day, he was with a colonel when they were approached by resistance fighters. The French had just found two collaborators and said they couldn't let them go because they were informers. So the collaborators were handed over and the resistance disappeared. The colonel turned to Frank's father and said, 'Look, we can't take them prisoner, we don't have the capacity. Take them round the back of that barn and finish them off.' It was an order – an illegal order – but an order. So Frank's father walked them around the barn. They begged for their lives. He said, 'You have to run and run now, and if you look back I'm going to shoot you.' He started screaming at them. 'Run! Fucking run!' And they did. They ran. He thought he might be able to shoot them in the back. But he couldn't do it. He let them go.

Frank never doubted he'd do the same, that he'd disobey the order, that he wouldn't shoot them. And he still thinks that. He *knows* that. But this place, SIZO-1, is giving him a lesson in the power of fear. Right now he's not sure what he'd do to get out of this place, but he thinks he might do things he never thought possible. Because he's scared. Right now, in this place, he's scared.

Frank Hewetson's diary

13th October
The sign of things to come. Snow has fallen and covered the Pig Pen roof, the loudspeaker and the watchtower. It's going to get much colder.

In countries across the world the demonstrations against Gazprom are continuing. Protesters picket the Albertina museum in Vienna where an exhibition sponsored by the oil giant has been inaugurated by the curators – with great solemnity – as the *Gazprom Collection*. In Paris an anti-Gazprom protester hangs for a day in a tent suspended from a rope tied to the underside of the Eiffel Tower. At the Barcolana Autumn Cup Regatta, a yacht race in the Gulf of Trieste in Italy sponsored by Gazprom, protesters in a RHIB hold banners calling for Arctic protection as they speed alongside the company's entrant – the 35-metre *Esimit Europa,* emblazoned in Gazprom blue, the sail rendered as one huge logo. The declared mission of the Russian yacht is to 'unite Europe' but images of security guards in Gazprom-branded tracksuits trying to stab the Greenpeace boat with a knife succeed only in uniting opinion against the company.[78]

The mother of Gizem Akhan, a 24-year-old activist from Turkey now jailed in SIZO-1, writes an open letter calling on more people to join the movement to free her daughter. 'They all took a step for the future of this Earth, marched on and stood up for something. We must stand beside them. They must not feel alone.' Her plea is echoed by the Mothers of the Plaza de Mayo[79] – the association of Argentine women whose children were among the thirty thousand 'disappeared' between 1976 and 1983 by the country's military junta. The same day Archbishop Desmond Tutu writes to the Russian government calling for the release of the Arctic 30. 'This is not just an Arctic or a Russian issue,' he says. 'The impacts of global warming will be most keenly felt in the developing world.'[80]

Also that day protesters from the pro-Kremlin youth organisation *Nashi* picket the Greenpeace office in Moscow. *Nashi* – meaning 'Ours!' – numbers tens of thousands of 17–25-year-olds who are dedicated to defending the Putin regime. Its members meet at summer camps across Russia to receive basic military training. The group has been accused of harassing opponents of the Kremlin.

Today Putin's young supporters are carrying banners depicting the faces of famous Russian Arctic researchers above the slogan 'REAL DEFENDERS OF THE ARCTIC!' The campaigners from the Dance Hall invite the *Nashi* protesters to come inside the building to discuss their issues. The Putin loyalists appear uncertain, then agree to the meeting. Inside the office they drink coffee with Greenpeace campaigners who tell them Russia's Arctic researchers are certainly heroes, but so are the activists in jail for trying to protect the frozen north. The *Nashi* protesters say little, finish their biscuits and leave.

The next day eleven Nobel Peace Prize laureates write to Putin demanding the release of the thirty, describing the Arctic as a 'precious treasure of humanity'.[81] They are later joined by two more Peace Prize winners – Lech Wałęsa and Aung San Suu Kyi.[82] Reuters reports that Paolo Scaroni, the CEO of Italian oil giant Eni, which has a multi-billion-euro partnership deal with Gazprom, has asked the Russian company to intervene to free the Arctic 30.[83]

In London John Sauven takes a call from New York. Paul McCartney has heard about the campaign and he wants to help. Putin is a huge Beatles fan, he sat in the front row when McCartney played 'Back in the USSR' in Red Square. The Moscow office has been saying for weeks that Paul McCartney is the only person who can get more media coverage than Putin in Russia.

And they're right. McCartney writes to Putin, and when the letter becomes public the last few paragraphs are printed in newspapers across the country.

Forty-five years ago I wrote a song about Russia for the White Album, back when it wasn't fashionable for English people to say nice things about your country. That song had one of my favourite Beatles lines in it: 'Been away so long I hardly knew the place, gee it's good to be back home.'

Could you make that come true for the Greenpeace prisoners?

Sincerely yours

Paul

TWENTY

Potatoes are bouncing off the walls like silver balls in a pinball machine. Sini is crouched in front of the window with a bag of them between her feet, occasionally glancing back nervously at the cell door then reaching down and selecting the smoothest and roundest of the hoarded vegetables.

The window is dirty glass, set back with a row of thick bars in front of it, and the only way to open it is to pull the ventilation lever which makes a narrow slit open sideways on. She thought she might be able to drop potatoes through that gap, but she can't reach it. Instead Sini is bouncing them against the wall, trying to make them rebound at such an angle that they'll ricochet through the gap and tumble into the courtyard below. But it's impossible. Most of the potatoes are hideously shaped, they've been boiled then left to go cold so they have a springy quality that makes their flight utterly unpredictable. They rebound off the wall at obscene angles, sometimes hitting her in the face or landing behind her on the bunk.

For hours she bounces potatoes. But she can't get it right. They won't go through the gap. She leans back against the wall, slides down onto the floor and buries her head between her knees.

Frank Hewetson's diary

19th October

Most of my GP goodies bag has gone between the 3 of us. Bit miffed the cheese went so quickly as I was quite keen on that. Still, it's all share and share alike which keeps the cell on an even keel. One of the guards seems to be walking round with a massive wooden mallet. It's so big it's comical. Boris + I get the giggles when we see him. At 18:30 tonight will be exactly 1 calendar month since arrest by FSB Spetsnaz.

Just made a birthday card for Boris. He turned 22 today. He's up for manslaughter and looking at 8–12 years. It's one of the saddest most forlorn cards I think I've ever written. There was not a single card or letter for him all day. He is never ever getting out of this system. You can just tell it a mile off.

20th October

Reading a book on my bed (approx. 5pm) when I heard Phil's voice very loud through the hatch of his cell door. He was ranting about being promised his phone call and having waited 3 weeks for it. I could just make out them fobbing him off so I slammed the door of 320 and shouted for them to give him his feckin phone call. He sounded really really upset so I started really slamming the door. The whole hall reverberated. Eventually Phil swore at them saying no one understood a word he was asking. I think he's at a really low point right now. 3 really young kids, his dad recently died. Poor fucker.

21st October
Still trying to establish what that huge mallet is for. The
guards turn up each morning with it but God knows how
you'd have room to swing it in a cell. I'm still greeting them
with a brisk Good Morning when the cell door opens and
we get marched outside for inspection. Yuri says I'm starting
to look thin and a bit gaunt. Well there's a surprise! I like
him though. He has got a real spark of life in him. He was
so dumbfounded when I beat him at chess last night.

By now all the crew have radios and televisions in their cells.
The radio plays soft rock classics, nothing to stir the blood,
but the jingle soon feels familiar, even comforting – a glock-
enspiel playing the first few bars of 'Midnight in Moscow' by
Kenny Ball. On their TV sets they watch euronews and the
music channels, and sometimes Russian state TV.

There are three kinds of programmes. Documentaries
about the FSB raiding someone, films about the FSB raiding
someone, and soap operas about the FSB raiding someone.
Frank sits for hours staring at the screen, flicking from channel
to channel, watching uniformed officers bashing down doors
in towns and cities across Russia, just like a young Vladimir
Putin once did. He sees what's happening here. It's being rein-
forced all the time, how there are bad people out there and
the state will protect you.

Camila is watching euronews when she sees video footage
of the protest at the *Prirazlomnaya*. The voiceover is in
Russian and she can hear the word '*hooliganski*' repeated over
and over. Then, rolling across the bottom of the screen in
English, she sees the news ticker.

PIRACY CHARGES DROPPED

Camila's heart jumps. She grabs her spoon and drops onto her knees in front of the radiator pipe. In the cell next door Alex is woken by manic tapping. Sini hears it too.

> *Camila: turn on euronews*
> *Alex: why?*
> *Camila: turn on euronews*
> *Alex: wow*
> *Camila: piracy dropped*
> *Alex: means they can't keep us in here*
> *Camila: amazing*
> *Alex: no reason to keep us in here*
> *Sini: careful don't get hopes up now charged with hooliganism*
> *Alex: sure?*
> *Sini: hooliganski means hooligan*
> *Alex: they can't hold us in here for that*
> *Sini: they can*
> *Alex: not as bad as piracy*
> *Sini: could be worse*

For a moment Alex and Camila thought they were getting out, but soon enough Sini convinces them this could actually be worse. They're moving from the pantomime charge of piracy to something that might stand up in the court of global public opinion. Alex wraps herself in her purple ski jacket and hugs herself.

Across SIZO-1 the activists are reaching the same conclusion. That night the road buzzes with discussion and analysis. Hooliganism carries seven years in jail. Nearly everybody thinks the change of charge is bad news. They fear they're being set up to get screwed by the Russian courts. Phil sits down on his bunk and writes an open letter, and the next morning he hands it to Mr Babinski.

Why I'm not a hooligan, by Phil Ball, aged 42½ …

Now we, the 'Arctic 30', face the charge of hooliganism. At first, it sounds only a bit more serious than naughty rascal or cheeky monkey. Something must be lost in the translation because seven years in prison seems a bit harsh. My sons, aged seven and nine, will be teenagers and my little girl will have forgotten who I am if I get out of here in seven years. Pretty unfunny too.

The small print of the charge says that hooliganism is a 'gross violation of public order' and 'in contempt of society'. Well, hang on just a moment: 'contempt of society'? I give blood; volunteer at my local scout group; pick up dog poo off the playing field and I don't have a dog; went to court as a witness to two violent crimes; helped fight a supermarket development; have taught kids to make award-winning films; have worked on projects for the Stop Aids foundation and the RSPB [Royal Society for the Protection of Birds]; invested £1,000 of my own money to help set up a community wind farm co-operative; and once saved and hand-reared a pigeon called Gerald. But the biggest thing I've done in support and protection of society? Coming 180 nautical miles north of the Arctic Circle to protest against Arctic oil drilling, against the greedy mega-rich oil companies Gazprom, Shell and others that do not listen to the warnings about oil spills, runaway climate change, hurricanes, droughts, floods and famines, and continue to make a fortune at the expense of and 'in contempt of' the societies of our children and grandchildren.

Hooliganism doesn't even come close to what they are guilty of. So, no, I'm not a pirate and I'm not a hooligan. OK? Can I come home now?

Phil Ball — father of three cheeky monkeys and one of the Arctic 30

Frank Hewetson's diary

24th October

Just saw local news item informing that piracy law has been dropped but replaced with hooliganism law that covers 0–7 years as opposed to 10–15 for piracy. Shortly after that our cell had a visit from 9 officers, 1 interpreter and a prosecutor. They claimed my pictures of Nell, Joe, Pluto and Free Frank protest contravened several sections of the prison code as they had been stuck on the wall by my bed. There was a tense stand-off as they repeated the infringements and I was surrounded by 9 screws with Boris and Yuri standing behind me. It was a potentially dangerous moment where I could feel my emotions rising fast. I eventually bowed my head and agreed. They slowly walked out one by one without saying a word. It was awful. I looked back up at Nell and Joe and slowly climbed onto my bunk and removed them with as much care as possible to avoid damage. It was quiet. Boris and Yuri looked down at the floor. Then thank God Yuri looked at me very closely and said, 'Frank, game chess??' He read the situation perfectly. Not forgetting that he is looking at 5–7 years inside.

Dima has visitors. Two men in sharp suits, not the cheap acrylic type worn by the investigators at the local FSB head-quarters in town.

They're standing in the middle of his cell, examining the contents of his shelves and the bags of food piled up in the corner, at the peeling green paint and the small barred window. Dima is watching them from his bunk. He doesn't know who they are, but he's been in jail long enough to know they're trouble. Thirty seconds ago they opened the door and walked in unannounced. No greetings and no orders. Instead they merely closed the cell door and each took a step forward then put their hands on their hips.

Both guys are in their mid to late thirties. One is spindly thin. He has greasy hair that's parted in the middle and sticks to his forehead, small red shaving spots speckle his neck, two deep crevices cut their way from his nostrils to his jaw, above which stand two bony cheeks that collapse into his mouth. Surrounding his thin lips are a few wispy whiskers – a failed attempt at a goatee beard, perhaps? He looks a little like a giant emaciated gerbil. The other guy is carrying some weight, sandpaper stubble, a thick helmet of brown hair that's cut along a savagely straight fringe.

The men ignore the Russian prisoners and walk over to Dima's bed. Gerbil kicks the leg of the bunk. It makes a metallic rattle.

'Dimitri Litvinov?'

Dima sits up. 'Can I help you?'

'Come with us, please.'

Dima rubs his greying beard and eyes the men suspiciously. Then he swings his legs over and jumps down.

He's taken from his cell to a small room containing a desk, three chairs and a huge portrait of Putin. One of the men locks the door, Dima is told to take a seat. The men sit opposite him. Dima's heart is beating fast now. They're quite obviously senior FSB officers, probably up from St Petersburg. Today has already seen a march in Moscow demanding the release of political prisoners. Twenty thousand attended – a huge number for a protest in Russia, where demonstrators can be plucked off the street and thrown in jail by a political police force empowered to act with impunity in defence of the President's agenda. One of the columns on the march was dedicated to imprisoned environmentalists, with the focus on the Arctic 30. There was a fleeting, dismissive reference to the protest on the TV news. Is this what the men want to talk about?

Gerbil crosses his legs, explores the inside of his mouth with his tongue, lifts an eyebrow in a way that says, *Well, here we are then*.

'How are you guys doing?' says Dima nervously.

Gerbil nods slowly. 'Good, Dimitri, good. How are you?'

'I've been better.'

'Yeeees, I can imagine.' The man leans forward, sniffs, meshes his fingers together to form a two-handed fist. 'Listen, now the piracy charge has been dropped we thought it was time we started talking. You and us.'

'Okay.'

'It's getting serious now, Dimitri. We're in the end game. The hooligan charge is going to stick, you're looking at seven years, you know that. So we'd really like to hear from you, this time without any protocol' – he means Article 51 of the constitution, the right to silence – 'and it would be really good if you gave us some answers. To the questions, I mean. You know, about what happened out there.' He waves a hand in the air. 'We know most of it anyway, but just to get it confirmed. It's your only chance of skirting that rap.'

'Okaaay.' Dima draws a deep breath. 'And who are you, exactly?'

The man clears his throat. 'Well … ' He smiles. 'We're the competent authorities.'

'The competent authorities?'

'That's right.'

'And what is your field of competence, exactly?'

His bottom lip protrudes for a moment as he shrugs. 'This.'

'And … what is *this*?'

'Don't you understand what I'm saying to you?'

'I'm afraid I don't.'

'We're the competent authorities.'

'You already said that.'

'Dimitri, please, come now, stop messing around, stop playing these … ' his face creases, he touches his nose ' … these *childish* games. Let's just get the questions answered, okay? Better for everyone.'

'I'm perfectly willing to answer any questions from the investigators.'

'Ah, good.'

'With my lawyer present of course.'

'Aaaaah, the lawyers.'

'I'm afraid so.'

'Come come, don't you understand what these lawyers are after? They want to keep you in prison for as long as possible, you know that, Dimitri. Paid by the hour. Making a *lot* of money.' Gerbil taps the edge of the table with a finger. 'How much do you make, Dimitri? At Greenpeace?'

'It's not about the money.'

'Sure.' He nods with faux sincerity. 'Sure it's not.'

'Really.'

'Well I'll tell you, they make a lot more money than you do, those lawyers you've hired. A *lot* more. They really don't want to get you guys out, not while they're raking it in. Nooooo. So look, let's forget about the lawyers for a moment and let's talk about you, about what's in *your* interests. We can make your stay here much more comfortable, a lot more comfortable than it is now. And all you have to do is answer our questions.' He looks away. 'It's just for background anyway.'

'My stay in prison is perfectly comfortable already, thank you.'

'Well, you know, it could be much worse. Think about that, Dimitri.'

Gerbil pushes the chair back and stands up. His colleague does the same.

'That's it?' Dima asks.

'For now.'

The other cop unlocks the door. Gerbil tugs Dima's arm and leads him out. Putin looks down, silent, inscrutable.

TWENTY-ONE

Dima is lying on his bunk, smoking a cigarette, running over what happened back there. Is he being singled out for special treatment? Is he going down for years while everyone else gets released? Nobody else is getting this shit, or if they are then they're not talking about it on the road. For hours Dima's been reconstructing the conversation with the competent authorities, but he can't work out what it means, and now he's exhausted.

He sucks on the butt of his cigarette, crushes it against the inside of a tin can and lights up another.

Dima Litvinov never thought he'd become a real smoker, but honestly, there's no point *not* smoking in here. Vitaly and Alexei both smoke strong Russian cigarettes all day long and a thick fog hangs in the cell whenever somebody is awake. Sometimes Dima can barely see the opposite wall. Just by being in here he inhales as much smoke as he'd ever get from actually smoking himself. So sometime around the second week he decided he might as well get some pleasure from it. He took up cigarettes.

They come free on the *doroga*, all you do is send a note out and the *kotlovaya* will arrange for a packet to be sent your way. And once he started, he realised there was no point in giving up. He could have quit but he'd still suffer the same harm to his health, and without getting any of the pleasure. Cigarettes give a rhythm to the day. They break the boredom. The only way it would be worthwhile quitting smoking in this cell is if he could get the other two to quit as well, but there's no chance of them giving up because the rest of the cell is always smoking.

Dima takes a drag and stares at the ceiling. It's a classic paradox. It only takes one of them to light up to make it pointless for the others to give up, because they share the same air. Maybe I should just quit, he thinks. Then persuade the others to give up too. Maybe somebody just needs to jump first.

The women tap to each other in code. Frank has his Valium prescription. Dima smokes cigarettes or goes *uyti v tryapki* – 'into the rags'. Each of the thirty finds a different way to survive. For Anthony Perrett, it's the *Gulag Chronicle*.

The idea came one night soon after they were jailed, when he was talking to Phil on the road. Both love to draw, from day one they immersed themselves in sketches of the cells, of the other prisoners, the guards, the raid on the ship. And it's not like there was a shortage of news. So Phil suggested they start a newspaper. His plan was to circulate a daily with empty space where the other prisoners could write their own stories. Anthony was less interested in what the others had to say, instead he wanted to write, edit, illustrate, publish and circulate his own paper. Phil pushed ahead with his idea and for two days his publication – the *Gulag Gazette* – had a tabloid monopoly. But Anthony was merely biding his time, for he had an altogether different concept in publishing.

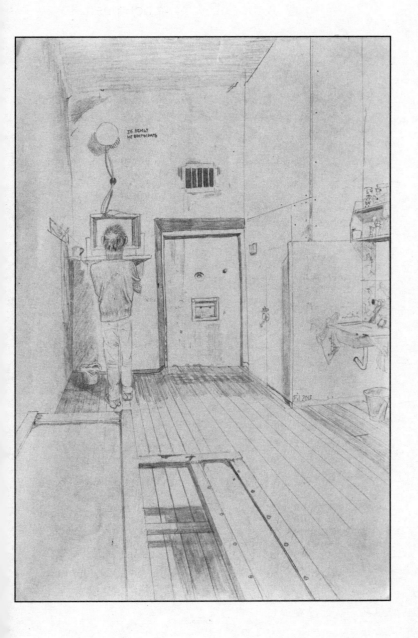

Window end of Cell 313
Pirl of the Arctic

The *Gulag Chronicle*.

He launched with three editions on a single night, including an editorial that ripped into Phil and his dirty rag. 'Do not read the Gazette, it's nonsense, the editor's a moron.' Then he sent it all out on the road.

Soon enough the *Gazette* folded and demands flowed into Anthony's cell to make the *Chronicle* into a daily. It became a prison fixture. His serving of satire and gossip was eagerly anticipated by the activists. The sixteenth edition was a particular hit with the readers. On the front page, in banner type, sat the headline: 'SIT STILL AND SAVE THE WORLD'. And below that ran the day's lead story.

It has been observed that we merry few locked in our cells are, willingly or not, completing many campaign objectives by sitting on our arses thinking of home. Despite our jailors, we are doing our jobs better the longer our incarceration continues. Allow a small prophecy if you will. This detention shows many nations that Greenpeace will not cower in the face of adversity, it will persevere where others fear to tread. This persistence will put negative press at the top of all oil companies' board meeting agendas. These risk-averse project-monkeys will be forced to look at new energy sources, which will eventually lead to new economies and end the march into the northern wilderness for black gold!

Five weeks into their ordeal, with their story making front pages across the globe, the *Chronicle* has become the Arctic 30's own in-house newspaper. Every evening the activists

THE GULAG CHRONICLE

16th Edition

5th October 2013

SIT STILL AND SAVE THE WORLD

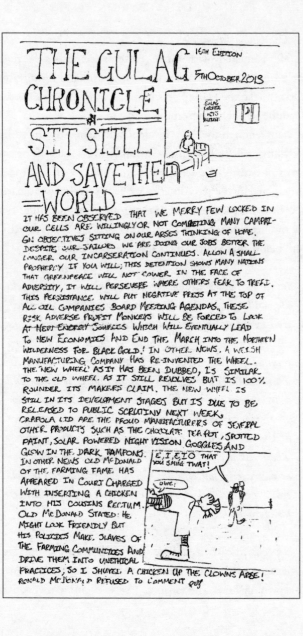

IT HAS BEEN OBSERVED THAT WE MERRY FEW LOCKED IN OUR CELLS ARE, WILLINGLY OR NOT COMPLETING MANY CAMPAIGN OBJECTIVES SITTING ON OUR ARSES THINKING OF HOME. DESPITE OUR JAILORS WE ARE DOING OUR JOBS BETTER THE LONGER OUR INCARCERATION CONTINUES. ALLOW A SMALL PROPHECY IF YOU WILL; THIS DETENTION SHOWS MANY NATIONS THAT GREENPEACE WILL NOT COWER IN THE FACE OF ADVERSITY, IT WILL PERSEVERE WHERE OTHERS FEAR TO TREAD. THIS PERSISTENCE WILL PUT NEGATIVE PRESS AT THE TOP OF ALL OIL COMPANIES BOARD MEETING AGENDAS, THESE RISK ADVERSE PROFIT MONKEYS WILL BE FORCED TO LOOK AT NEW ENERGY SOURCES WHICH WILL EVENTUALLY LEAD TO NEW ECONOMIES AND END THE MARCH INTO THE NORTHERN WILDERNESS FOR BLACK GOLD! IN OTHER NEWS. A WELSH MANUFACTURING COMPANY HAS RE-INVENTED THE WHEEL. THE "NEW WHEEL" AS IT HAS BEEN DUBBED, IS SIMILAR TO THE OLD WHEEL AS IT STILL REVOLVES BUT IS 100% ROUNDER ITS MAKERS CLAIM. THE NEW WHEEL IS STILL IN ITS DEVELOPMENT STAGES BUT IS DUE TO BE RELEASED TO PUBLIC SCRUTINY NEXT WEEK, CRAPOLA LTD ARE THE PROUD MANUFACTURERS OF SEVERAL OTHER PRODUCTS SUCH AS THE CHOCOLATE TEA POT, SPOTTED PAINT, SOLAR POWERED NIGHT VISION GOGGLES AND GLOW IN THE DARK TAMPONS. IN OTHER NEWS OLD MCDONALD OF THE FARMING FAME HAS APPEARED IN COURT CHARGED WITH INSERTING A CHICKEN INTO HIS COUSINS RECTUM. OLD MCDONALD STATED: HE MIGHT LOOK FRIENDLY BUT HIS POLICIES MAKE SLAVES OF THE FARMING COMMUNITIES AND DRIVE THEM INTO UNETHICAL PRACTICES, SO I SHUVEL A CHICKEN UP THE CLOWNS ARSE! RONALD MCDONALD REFUSED TO COMMENT ouff

STOCKS $ SHARES

FRANK = 320   DIMA - 423
TOMEK = 322
PHIL = 313
HERNAN = 309
DENIS = 314
KIERAN = 308
ANT = 311

LETTERS

HEY DIMA,

COULD YOU WRAP THIS AND SEND IT
SPECIAL DELIVERY TO TOMEK WHO
WILL TRY AND GET IT TO THE OTHERS

CHEERS BUDDY
A

Hey to everybody,
Am I the only one who got was a shitty
package again? If not let's ask the office
to buy better quality staff things. I can
pay myself but for god's sake it's the only
nice suprise here :) ~~~~~~~~~~~~~~~
Actually the last one was even worse than
the first which was really disappointing,
considering very bad food here. All the best, Tomek
07. 10. 2013

PS. I got 1 chocolate, some instant
coffe, some black tea ▓▓
4 apples, sugar,
a little oil, and
some clothes.
(underwear)

GULAGTADA - 2013

MURMANSK - SOCHI

would read the *Chronicle*, tick off their names to say it's passed through their cell, then pass it on. There's a comments box in the corner – letters to the editor – and a stocks and shares section that gives updates on who's in which cell. *Frank = 320* or *Phil = 313*.

It takes Anthony hours to write it and illustrate the pages. There are stories about the *Prirazlomnaya*, about helicopters, a balaclava sale. One story details the invention of the colour invisible by the artist Lucian Fraud. He illustrates it by cutting a hole in the paper where a drawing would normally be. But the *Chronicle* is mostly rants against oppression. Anthony worries that someone will get caught with a copy that castigates Popov and the regime. So he starts writing the *Chronicle* in such a way that if you speak English as a first language you'll understand the double-entendres and the sarcasm – 'Popov's moustache has no fascist connotations and is indeed one of the superior facial growths of the faculty' – but it won't get anybody into too much trouble if the authorities find a copy.

Sini reaches up and rubs her forehead. It's just been struck by a rebounding potato.

She's crouched in front of the window, the ventilation gap is opened sideways on, potatoes are bouncing everywhere except through that narrow gap. For days she's been trying it, but the more frustrated she becomes the less likely it seems that one of them will bounce through the window.

She's ready to give up now. She's ready to accept that Popov will discover her secret stash of contraband potatoes and send her to the punishment cell. She'll survive it, she knows that.

She's strong, stronger than she was when she arrived here. She's resigned to it now, ready for a spell in the cooler. She just hopes she doesn't lose this cell, with Alex and Camila just along the corridor. She takes another potato, throws it in the air and catches it. Then she flicks it and watches it bounce off the wall and fly clean through the gap and disappear into the void beyond.

She jumps up and peers down through the window.

Wow, I did one!

Okay, so maybe it's not time to give up just yet. She crouches down and tries again. And again. And half an hour later another one tumbles out of her cell. She stops and considers what she's been doing differently and realises that up until now she's been throwing too hard, she's been putting all of her frustration, her anger at Popov and Putin, into those potatoes. But when she breathes deeply and relaxes, when she puts less force into the throw, then the flight of the potato is more easily controlled. So she starts holding them between thumb and forefinger, adopting a more subtle, more delicate launch style. And by doing this she finds she can land every tenth potato, then maybe every fifth one, until her entire potato mountain finds its way through the gap.

Every day Sini gets nine new potatoes, she eats four and in the night she gets rid of the rest. The stockpile is gone now. Her cell smells normal. The threat from Popov has tumbled into the courtyard below.

One of the guards is tugging at Frank's T-shirt, but Frank is asleep so it takes him a few seconds to realise that somebody's pulling him out of bed, but he can't bawl out this person because it's a guard and he's in a Russian prison cell. New day, same reality. By the time he's been hauled from his bunk, Frank is awake. And by the time he's handcuffed behind his back and is being marched down the corridor, he is very awake indeed.

He's taken down the stairs and along various unfamiliar hallways until he's outside a heavy door. It swings open, he's pushed inside, and before him, sat behind a broad desk, tapping at a computer keyboard, his red face betraying extreme dissatisfaction, is prison governor Popov. His mouth is tight with rage, there are little bubbles of spittle on his lips. Standing behind him is a woman, the prison translator.

Popov looks up. He eyes Frank closely but says nothing. The tip of his tongue runs along his upper lip. As he slips it inside his mouth there's a flash of gold. Frank clears his throat and starts to say something, but Popov presses a finger to his lips to demand silence. Then he grips the side of the computer screen and turns it around so it's facing Frank. And there, blown up large, is a photograph of Frank behind the bars of a courtroom cage, over the headline: 'I MIGHT PUT IN A COMPLAINT TO THIS HOSTEL'S MANAGEMENT'.

A week ago Frank used Mr Babinski to smuggle a letter out to his friend Lisa. The letter was written in the style of a TripAdvisor review. He awarded SIZO-1 zero stars:

Since arriving at and being processed through the system at Murmansk State ~~Prison~~ Hostel I have been keeping a mental and digestive diary of food substances supplied through the security hatch of our cell door, three times a day.

Breakfast, 06:00 *It looks like porridge … It is porridge.*

Lunch/abyet, 13:00 *Potato is in there somewhere. The trick is to sieve out the suspected meat particles and positive ID them before consumption. It is often quite wise not to consume them.*

Dinner/oozhin, 19:00 *Potato makes a comeback on most evenings, indeed lunch makes a comeback on some evenings, but that's quite often well before 19:00. If one gets extra boiled water and uses bread, dinner can actually be quite palatable.*

Once inside the accommodation one can really appreciate the protection and enveloping sensation of a 5m x 2m cubicle. I think these are somewhat larger than the variety I have stayed in at Shinjuku station and downtown Osaka but some of the features differ in quite remarkable fashion.
I might try to write a letter to the management and leave it in the 'comments' box attached to the 5 inch plate steel partition in the hallway.

Frank's friend Lisa, to whom he sent the review, is the editor of the *Independent on Sunday*, and now it's on the front page of that day's newspaper – a UK national title owned by the Lebedevs, a Moscow family which also owns 49 per cent of the liberal Russian newspaper *Novaya Gazeta*. Frank's review of Popov's regime is being read across Europe and Russia.

Frank looks at the screen and sucks his teeth. 'Ah. Right. The food review.'

'Yeeees!'

'Hmmm.'

Popov says something to the translator, then he launches a furious diatribe in Russian, with the translator racing to keep up.

'How could you do this? We give you all nice things here, we make it nice, but you do this to me. To me! You say these lies, these *lies* about nice food you get. You come here and think Russia is shit, but no, you are shit ... you ... *you* are shit with these lies you tell.' Popov is spitting out the words like he's firing them from an AK-47, a *rat-a-tat-tat* of abuse directed at Frank, one hand clenched into a fist that's banging on the table for emphasis, the other with a long extended finger that's jabbing at the screen in time with the tirade. After a minute, maybe two, Popov reaches a crescendo of abuse, falls back in his seat, throws his hands in the air, spurts another few words – 'All lies you tell, typical Western lies!' – then he folds his arms and falls silent.

Frank runs his hands through his blond hair, which by now has grown out completely. 'Yes, well ... but you see, that letter, it wasn't *actually* from me.'

'Uh?'

'It's not me.' He shakes his head. 'Not me.'

'*Is you!*'

'Oh, come on.'

'*Whaaaat?*'

'You don't seriously believe everything you read in the *Independent*?'

'Uh?'

'It's. Not. Me.'

Popov sucks a long frustrated breath through his nose then, with a stiff outstretched arm, he points at the door. Frank is dragged from his seat and marched out of the office. And in the following days the Arctic 30 notice a marked change in the guards' attitude to contact with the outside world – with lawyers, consuls, the ground team and human rights observers. The authorities are instituting a clampdown on people bringing items to and from SIZO-1. Even Mr Babinski struggles to conduct his vital work.

Popov is taking his revenge.

TWENTY-TWO

'Ha! Evidence of illegal inter-cell communication!'

'What?'

The guard shakes his head slowly then leans forward so Dima can feel his breath on his face. 'These letters are a clear breach of regulations.'

'Aww come on, this is crazy.'

'Don't you "come on" me. You'll have to answer for this, Litvinov.'

'But they're just letters.'

'Illegal communications.'

'Seriously?'

Sheets of white paper crunch as the guard squeezes them in his fist. A smile breaks on his lips.

'Yes. Seriously.'

Each day after breakfast the cells are subjected to a rudimentary search, but every few weeks the guards sweep through the prison pulling the place apart. It's called a 'deep search'. The prisoners take everything they own out into the corridor and pile it up against the wall. The guards then go through the cell checking every surface, under the bunks, behind the toilet, everywhere. They take a huge mallet and whack the metal frames of the bunks, then they strike each of the bars at the window. They're listening for a solid resonating ring – evidence that the metal has not be sawn through. They don't want detainees arming themselves with metal piping or cutting through the bars. Then the guards go through the pile in the corridor. Rope from the *doroga* is confiscated, maybe the *domovaya* is ripped up, a copy of the *Gulag Chronicle* is examined by a confused guard before being dropped into his bag to be burned later. And it's in the course of one of these swoops that a guard has pulled four pieces of paper from the inside pocket of Dima's jacket, on which he has written drafts of letters. One is to his wife Anitta, the others are to colleagues and friends.

Dima holds out his hand. 'Come on, don't be silly. Give them back.'

'We'll be taking these as evidence, thank you. Illegal communication. Illicit inter-cell messaging.'

'Look at who those are written to. That one, it's addressed to Anitta Litvinov. You're saying you have my wife in here too?'

'Oooh, so you were intending to send letters to your wife? Illegally!'

'Well, I'm allowed to put it in the envelope and send it out that way, right? I can post it out, yes? That's not illegal.'

'Is that what you were going to do?'

'Yes.'

'Bullshit.'

Dima wonders if this is all part of the crackdown on Mr Babinski since Frank's review of Popov's food. But he doesn't ask. He's careful not to reveal there's a secret system for smuggling out letters.

'Yes, I was going to post these. I was going to give these to you guys and have them posted out. What's so strange about that? And anyway, these are drafts.'

'Oh, really?' The guard unfolds a letter.

'Hey man! Maybe I don't want you to read what I've written to my wife!'

'I thought you were going to submit these to the censor?'

'Those are drafts. What, I can't write a draft letter without you guys reading it?'

'You're allowed to write a draft, but you're not allowed to send it out by illicit means. I now intend to have these translated. If they are what you say they are, you'll get them back tomorrow.'

And with that, the guard leaves.

The next day, nothing. The letters aren't returned. Dima writes a protest note and drops it into the complaints box. The following day he has a visitor, a representative from Popov's office.

'We looked through your letters. It seems obvious that you were planning to distribute them illegally, bypassing prison censorship. We will not be returning them. Come with me.'

'Where am I going?'

'To see the psychologist.'

'Oh, Jesus. Really?'

He's led down a corridor to the office of the counsellor. The man is on his feet in full camouflage, fingering his baton. The peak of his military cap is pulled low over his face so it nearly covers the coal-black lenses of his Reactolite glasses. He examines Dima for a moment then lowers himself into his deep leather-backed swivel armchair and points at the seat opposite. Dima sits down and takes a moment to consider what Freud might have said about a psychologist who works in full military fatigues and wields a weapon in the consulting room.

'How are you, Litvinov?'

'So so.'

'Any suicidal thoughts?'

'Nope.'

'Feeling depressed?'

'I'm not very happy about being locked up for something I didn't do.'

The psychologist nods. 'Sure, sure.'

'I was told you wanted to see me.'

'Mmmm.' He leans back. 'It's about these letters. They're, ummm … ' He scratches the corner of his mouth. 'They're talking about putting you into a punishment cell.'

'A punishment cell?'

'Three days in the *kartser*, because of these letters. They say you were going to send them out illegally. And, well, we can't have that.'

'But … but they were just drafts, I never even … the *kartser*? This is about those FSB guys, isn't it? The ones who pulled me out of my cell and threatened me. They ordered this.'

'I don't know about any of that. All I know is that the governor has asked me to make an evaluation, see if you're in a fit state for the punishment cell.'

'And am I?'

'You'll be fine.' He turns to the guards. 'Yup, he'll be fine.'

'That's it?'

'That's it.'

228

The guard takes Dima back to his cell and orders him to pack his possessions, everything he owns, including his bedding. As he fills his huge pink bag, Vitaly stomps up and down the cell. 'Whaaaat? They're putting you in the *kartser*? You have to have nine marks against you before you're sent to the *kartser*. They can't just do that. Why are they doing this?' His skin is dark but right now his cheeks are flushed red. He stops and addresses the guard standing in the doorway. 'Why are you doing this? This is crazy. And why does he have to take all his stuff?'

'Because he may not be coming back to the same cell afterwards. Orders of the governor.'

Dima is marched through the prison, up flights of stairs and down again, his bag slung over his shoulder. At the end of a long corridor the guard stops him and takes the bag. Dima is left standing in his T-shirt, sweat pants and slippers. Even his bowl and his cup are taken from him. The guard opens a heavy door and holds out his arm, inviting Dima to step inside.

It's tiny, almost bare. There's no bed, just a wooden bench on a hinge that's folded against the wall. High up near the ceiling is a window the size of a shoebox, too small to capture much of the Arctic sun. Dima steps inside, behind him the door swings closed.

It's quiet, dark, he's alone. He sits on the floor and stretches his legs, but they reach the other side of the cell before they're straight. Three days, Dima thinks. That's seventy-two hours. Minutes pass, then an hour, and another, or maybe it's been longer. Or maybe not. If he counts to three hundred that's five minutes, and if he does that twelve times then that's one hour. He gets to his feet and starts pacing back and forth, counting time. '… two-nine-seven, two-nine-eight, two-nine-nine, three hundred, one, two, three, four … '

Hours pass. He's still pacing when the door opens.

'Bed time,' says the guard.

'What time is it?'

'Ten o'clock.'

He's been here fourteen hours.

The guard unclips the bench; it drops down, the door closes. Dima takes off his steel spectacles and lays them on the floor under the bench. He sleeps for a while but wakes in the night, paces some more, lies down again, sits on the edge of the bench, stands up, sits down again. He can feel the early stages of panic coming on. He forces himself to think of Lev, his oldest son. Lev has been travelling the world this past year, sailing the South Pacific as a dive master on a three-mast schooner. Now he's en route to New Zealand to volunteer on an organic farm, but the last Dima heard he was on the island of Vanuatu.

Now he uses Vanuatu to suppress the fear. He stands back and looks through the window at the black sky, the wind from the Arctic whistling in. And he repeats the word. 'Vanuatu Vanuatu Vanuatu.' He's imagining coral atolls, palm trees and white sand. Lev splashing out to a canoe.

But soon enough his mind is running to a dark place. He can't make sense of this. Why has he been put here? It's definitely a bad sign. If they were preparing to release him then they wouldn't be doing this. His thoughts spiral down and down, down into the darkest place, the place where he's kept his worst fears locked away. And now, when he tries to climb out, when he tries to imagine those coral atolls and white beaches, they won't come to him, they might as well be on a different planet, because he's slipping into a quicksand of panic.

Turma racing.

This is it. This is how it's going to be now. Years of this shit, locked up in this hellhole of a prison, thrown into the kartser for daring to even look sideways at a guard. Seven years. How do you pace out that kind of time? How many seconds is that anyway? And what about Anitta? Shit, I'm going to have to tell her not to wait. I can't ask her to put her life on hold while I'm rotting away in this place. No, I'll tell her she's not to wait around for me to get out. She has to get on with her life. Fuck. Seven years. Seven fucking years.

His mind races and races, he paces again, sits down, lies down, stands up, looks up at the sky and strains for Vanuatu. He paces and paces until the guards return. It's 6 a.m. They fold up the bench and leave Dima a bowl of porridge. He ignores the food, paces the cell and counts to three hundred, and again, working through the minutes and the hours. The sun comes up and throws a bleak smudge of grey onto the wall for a few hours before retreating. Later the bench is lowered, another night, he sleeps and paces, lies down, stands up, sits on the bench.

I went to the Arctic to take on Gazprom, I thought it would give us a platform to talk about Arctic oil. I just never thought they'd keep us. But they did, and here I am. But that's okay. We were challenging Gazprom, and Gazprom is Putin, so of course I'm in a punishment cell in an isolation prison in the Russian Arctic. And maybe this is where I should be. If you really believe in something then you have to show you'll pay a price. What were we going to do, just hang a banner on that oil platform and say we'd done our job? We should be in jail. This is right. This shows we're winning. This is where I need to be right now.

The next morning, or maybe it's the afternoon, the door opens and a guard motions for him to step into the corridor. Dima rubs his eyes and looks up. Standing before him, extracting a piece of food from between his front teeth with the nail of his little finger, is Popov.

'Dimitri, hello.'

Dima's eyes narrow into slits. 'What time is it?'

'One o'clock.'

He's been in that cell for twenty-nine hours.

'Having a good time in there, are you?'

'What do you think?'

'Well *hopefully* this will teach you not to break our rules.' Popov jerks his head. 'Come with me.'

Dima puts on his spectacles. Popov leads him through the prison to the door of his office, then disappears inside, leaving Dima facing the wall, hands behind his back, a guard either side of him. Minutes pass. He wants to sit down, his feet are tired from pacing the *kartser*. Eventually a voice booms from the other side of the door.

'Okay, bring him in.'

It's huge, this office. Popov is sitting at a big desk behind a computer screen and keyboard. At the back of the room are two soft chairs facing each other below an enormous portrait of Putin. Popov starts speaking, and straight off his manner is oddly breezy. He uses the informal '*ty*' when he's addressing Dima, like they're old friends.

'You can take a seat,' he says, and when Dima is seated he extends an arm and shakes Dima's hand with vigour.

'Do you smoke?'

'I do now.'

'You started in prison?'

'Yeah. But I'm trying to quit.'

'Would you like a cigarette?'

'I'm in the *kartser*, I'm not allowed to smoke.'

Popov snorts. 'It's okay, don't worry.' He holds out a packet of cigarettes, Dima takes one, so does Popov. The governor lights it and breathes in deeply then two dark grey tusks appear as he exhales through his nostrils. He holds out the lighter for Dima then pops it back into his pocket, saying, 'Tell me, have you ever read Goethe's *Faust*?'

'I have.'

Popov takes a drag. 'So you know this whole thing about good and evil then?'

'Good and evil?'

'In *Faust*.'

'We all have good within us.'

'That's right Dimitri, yes. We're all fundamentally good.'

'And only errors of judgement make good people do bad things, but even bad people—'

Popov interrupts, saying, 'If a person keeps striving, Dimitri, even bad people, if they keep striving then their mistakes will bring them closer to righteousness.'

'So says *Faust*.'

'If they *strive,* Dimitri. If they strive to be good.'

'Yeah, I've read it.'

'And you must strive. All of you. What you did, it was a mistake, you know that. But you can be good people again, I know you can. And you, Dimitri' – he uses the familiar '*ty*' again – 'you are somebody who can strive, who *must* strive. It's in your blood, you're a Litvinov, and by recognising your errors of judgement you can become good again. Don't you think?'

Dima leans back and draws on his cigarette. 'It rather depends on whether you think holding a peaceful protest at an Arctic oil platform is an error of judgement. Some people, and I believe there are many millions of them, might say that us being kept here, in this prison, is a more fundamental error of judgement, and is one that says much more about the nature of good and evil than our climate change campaign.'

Popov berths his cigarette in the ashtray and holds his palms together, as if in reflective prayer. Then he taps his chin with the tips of his fingers, contemplating Dima's assertion, before leaning back in his chair. 'What do you understand by the concept of nationhood?'

'Russian nationhood?'

'Nationhood. Russian, American, whatever.'

'It's bullshit. I'm an internationalist.'

'Well I'm a nationalist, Dimitri. I'm not afraid of those words. Nooooo, I'm not afraid of saying I'm a nationalist, not at all. Not in the least bit. I'm a nationalist, and what you did, what you are doing, is a threat to my nation.' He lifts his shoulders, as if what he's saying is the most obvious thing in the world. 'It's really that simple. And I think that you're going to be feeling the wrath of this nation.' He's nodding slowly now. 'Because the nation is … ' He holds his hands out in front of him, like he's squeezing two invisible oranges, and his face creases as he searches for the words. ' … the nation is … is the *embodiment of the people*. And the state is the embodiment of the nation. So you see, the state, me, him' – his eyes turn to the giant portrait of Putin – 'are actually no more than the people. It is not me who has put you here, Dimitri. Don't you see that? It's not me, it's not the President. It's the people. The people have put you here.'

'The people?'

'Why yes, of course.'

'Well, if the people are so wise, why not ask them what they think? You could ask them all on the same day, and you could let them tell you secretly so they can't be intimidated, and you could let the media say what it wants in the weeks before this day. You could do all of that, and you could call it, I don't know, a *fair election*. Why not let them say what they will without censorship, so their wisdom can be appreciated by all of us?'

'Hmmm.' Popov stares over Dima's head into the middle distance. His eyes glaze, like he's suddenly absent from the conversation, then he blinks and almost to himself he mutters, 'History has been so unfair to the Gestapo.'

Dima's mouth drops open. 'The Gestapo?'

'Me, I respect your great-grandfather. He was close to Stalin, he knew the benefits of stability. Russia is a vast country, Dimitri. Our borders are hard to defend, our people are diverse, our languages many. Only through the primacy of the nation, embodied by the state, can we retain our place in the global order. But the state must be strong. Yes, the Gestapo … ' he smiles wistfully. 'We've learnt so much from them.' He makes a fist of his hand, raps his knuckles on the table and leans forward. 'Those guys knew how to run a prison. They stand as the master practitioners of penitentiary science and related systems. They're the ones who developed it all. All of it! Masters. Really, we owe everything to the Gestapo.' He sniffs. 'You look sceptical, Dimitri.'

Dima's not sceptical. He's furious. This guy's a clown, but he's also a thug. The Gestapo? The governor is everything that's wrong with Putin's Russia.

'Actually, I'm offended.'

'Because?'

'Because I hoped today's Russia wouldn't owe a debt to … to the fascists.'

'Fascists? What the hell kind of word is that? What do you mean, fascists?'

'Fascists. The Gestapo. They were fascists.'

'Ah, but Dimitri, what you don't understand is … ' And here Popov launches into a wider soliloquy on the nature of nationalism while Dima stares at his face, fuming at the man, watching the little moustache dancing on the upper lip, occasional flashes of gold from the capped tooth as the mouth spits out this cod philosophy. And all the time Dima's thinking, what does this man actually want from me? Why is he doing this? He's not asking me questions, he's just pouring all this out and I'm just sitting here cast in the role of student to Popov's master philosopher.

Eventually the governor runs out of steam. He tried to rationalise the contradiction between his admiration of the Gestapo and the immense pride he takes in the Soviet defeat of Nazism, but after several minutes he found himself in a verbal cul-de-sac before restating his argument with less conviction, and now Popov appears to have given up. The room is quiet but for the sound of a clock ticking. The fist in Dima's stomach is clenched tight and hard.

Popov breaks the silence.

'So, they're going to lock you up for seven years.'

Dima blows out his cheeks. 'Maybe. I don't know. I hope not.'

'Of course. Nobody wants to be stuck in prison for seven years. What are you going to do? What are you thinking? Because of course you're in charge of this situation, it's up to you. You want to be locked up for a long time? Is that what you want? What *do* you want, Dimitri?'

'Well, I've been waiting six weeks for you to give me a phone call.'

'A phone call?'

'A telephone call.'

'Who do you want to call?'

'Who am I gonna … I just want my call.'

'To speak to who?'

'I'm going to call my wife, of course. I'm gonna tell her not to wait for me.'

'What do you mean, not wait?'

'If it's seven years I don't want her to wait. I mean, seven years, it's too long. She should move on, find someone else.'

At this point Popov's nostrils flare, he grips the edge of the table and splutters, 'What? You can't do that!'

'Well, I can't have a woman wait for me for seven years.'

'No, no, no! The *family*, Dimitri. The family is the most important thing we have. No, you can't do that. Come on, they're going to be letting you out in two weeks' time, what are you talking about?' His eyes dart around the room until they fall onto one of the guards. He waves furiously, motioning for the man to step forward. 'You, yes you, make sure Dimitri sees the psychologist. He's becoming delusional, these things he's saying are extraordinary. He's not feeling well, he's … he's not himself.' He turns back to Dima. 'It's okay, you'll see the psychologist. He'll help you. Dear oh dear, telling your wife to go with another man. I'm afraid you're losing your mind, my friend.'

'I'm not. Really.'

Popov stubs out his cigarette and lights another. He examines Dima's face for a moment then leans forward and with great reverence he says, 'Tell me, have you ever read *The Red-haired Horse*?'

'*The Red-haired*—'

'Oh dear, Dimitri. My dear Dimitri, you *have* to read *The Red-haired Horse*.'

'Okay. It's a book?'

'About the Cossacks. True nationalists. Aaaah the Cossacks. I'm actually a Cossack myself.' He points at the guard. 'You. Do we have *The Red-haired Horse* in the library here? We do? Aaaah, very good. Okay, well make sure Dimitri has it in his cell tomorrow.' He turns back to Dima. 'You've got to read it. It's a great book.'

'Okay, yeah, sure.'

'Good.'

'Okay then.'

'Hmmm.'

'Yes.'

'It really is a great book.'

'Fantastic.'

Popov nods then takes a deep breath. 'Well, Dimitri, we can't chat like this all day. Back to the punishment cell for you.'

'Yup.'

Popov sucks on his cigarette and shrugs. The guard taps Dima on the shoulder, he gets to his feet and is led out of the office and back down the corridor towards the *kartser*. The knot in his stomach is tight, his heart is beating fast. This prison is run by a psychopath, he thinks, and I'm not sure if he loves me or hates me.

He's pushed into the punishment cell, the door closes, and he stands in the silence for a few minutes, confused, scared, alone. Then a key turns and the door opens.

'Come on, you're going back to the boss.'

'What?'

'He's not finished with you yet.'

Barely a quarter of an hour after leaving, Dima is sat back in the same chair, across the table from Popov.

'Cigarette?'

Dima nods. They spark up, drag deeply, exhale over each other's shoulder. Popov taps his cigarette over the ashtray and says, 'You know the FSB?'

'Yeah … I mean, of course I do.'

'Tricky guys.'

'Mmmm.'

'Oh *man*, those guys are tricky. You really should listen to them, though, do what they ask. That's the thing about co-operating with the investigators, it makes your life so much easier.'

'Well—'

'Don't mess with them, Dimitri. For your own good, don't be a hero. I know one of your guys is trying to be a hero, trying to take the blame for the attack on the platform. But really, what the fuck is he doing that for? I mean, everybody's already giving evidence anyway, he should just relax. Yes, everyone's singing now, telling the FSB who did what. No point in keeping schtum, eh? Everybody's singing anyway.'

'I'm not sure they are.'

'Oh they are. Yes yes, they are my friend.' He cocks his head to one side and sighs. 'But I can end this for you. You *do* know that?'

Dima scratches his cheek.

'Do you want me to end this for you, Dimitri?'

'What do you mean?'

'I can amnesty you from the *kartser* right now.'

'Okay.'

'So let's do it,' says Popov breezily. He opens a desk drawer and takes out a piece of paper and a pen. 'You write here that you're really sorry you broke the rules. You say it was your lawyer who forced you to do it, and we can all move on.'

'I can't write that.'

'Well okay, just write that you're really sorry, you'll never do it again and we'll date it so it looks like you wrote it yesterday.'

'That's all?'

'That's all.'

Dima thinks about it for a moment. He wonders if this is a trick, but by now he just wants to get away from this guy, so he scribbles down a few words – 'I'm sorry I broke the rules about letters' – and signs it, then spins the sheet of paper around and pushes it back to Popov. The governor claps his hands together and declares, 'Okay, that's it. Take him back to his regular cell.'

Dima twists his head and looks over his shoulder at the guard, like, is this guy for real? But the guard gives nothing away, instead he lays a hand on Dima's shoulder and a moment later he's being marched back down the corridor. His pink bag is handed to him, he swings it over his shoulder and he's taken back to his cell. Vitaly jumps up and throws his arms around him, but Dima breaks away. His heart is racing; the knot is like a rock in his stomach now.

'What happened? Dima, what happened to you?'

Dima hunches over a mug in the corner of the room, stuffs a fistful of tea into it, fills it with water and drops the immersion heater in. He lights a cigarette then turns to face his cellmate.

'Vitaly, you've heard of the Gestapo, right?'

'Sure.'

'And you know they were the bad guys?'

'Sure I do. Everyone knows the Gestapo were the bad guys.'

'Okay, good.'

And the next day a book is pushed through the feeding hatch in the cell door and drops onto the floor with a slap.

The Red-haired Horse.

Dima forces himself to read it that night. He's curious to know more about a work of such renown. It's set in the Soviet-era revolutionary period. Cossack traditions. Poorly written. The biggest piece of trash he's ever read.

TWENTY-THREE

Pete Willcox's diary

30th October

I guess I had better start a diary, to try and remember all this shit. Had another useless exercise period. Again it was drizzling and the exercise cell was half undercover, and the half that was still frozen was in the drizzle. So I stood in the corner and did the French Chairs. Better than nothing. At 1330 they said my lawyer was here so I went down to see Alexander. They had been waiting to see me since 10am. That really really sucks and it's just ball-breaking. I wonder if they are mad at Alexander for something? Anyway, no major changes, Alex confirmed they are trying to split us apart. That's not good. Alex also thinks they may cook up another charge for me. But he thinks even if they do, they will let me go with the other non-activists. I think he is bullshitting me and I should prepare for the worst. But realistically it is unlikely they will keep anyone in to Sochi. I heard Monday at the investigators that there is already talk of protests at the Olympics if we are still held. Not boycott, but protests. I do not think Putin is going to want that. Around 1630 they came to get Sasha [his cellmate]. I do not think it was because of me and I think it was because he did too much screaming. But the cell feels lonelier than usual tonight.

To understand Pete Willcox, you have to go back to his grandparents.[84]

In 1952 Henry Willcox chaired the US delegation to a peace conference in China. One of the speakers was the communist premier Mao Tse-tung. When they returned to America, Henry and his wife had their passports confiscated[85] at New York harbour. He sued the United States government, eventually securing a five-to-four Supreme Court decision on the 'right of freedom to travel'. But the case cost him his building company. He was voted out by the employees he'd given shares to.

The Korean War was raging, the Cold War was at its height, Senator Joe McCarthy was warning of 'reds under the bed' and the House Un-American Activities Committee (HUAC) was holding hearings that resulted in the famous 'Hollywood blacklist'. Three hundred writers and performers were boycotted by the movie studios based on their alleged communist sympathies. Even Charlie Chaplin.

On 6 March 1953, the day after Joseph Stalin died, Pete was born. His birth mother gave him up for adoption, but Henry's son Roger Willcox and his daughter-in-law Elsie gave the boy a safe, loving home. Two years later Elsie was herself summoned to appear before the HUAC to account for her suspected involvement with an anti-war group. Elsie was in the process of adopting another boy, Mike, but the papers weren't yet finalised. Fearing the summons could put the adoption in jeopardy, she took Pete and Mike and went underground for three months, hiding in a New England farmhouse. When the papers were eventually finalised, she surfaced and was issued with a subpoena to testify.[86]

Elsie was pulled before the Committee. One of the congressmen said she'd betrayed her country. Elsie invoked her right to silence. Pete Willcox grew up believing that if you hadn't been subpoenaed by the House Un-American Activities Committee then you hadn't done much with your life.[87]

He was five years old when he went on his first protest, against a new coal-burning power plant in Norwalk Harbor. Two years after that he was protesting outside Woolworth's in solidarity with the young African-Americans staging sit-ins at racially segregated stores in the Deep South.

In 1965 Pete witnessed the culmination of the famous march from Selma to Montgomery. Selma was the county seat of Dallas County, Alabama. In the early 1960s the population was 57 per cent African-American, but only 1 per cent were registered to vote[88] (for whites, voter registration sometimes surpassed 100 per cent[89]). A campaign began to turn that around, but it was met with organised violent resistance by sections of the white minority, and by the mid-sixties Selma had become a focus for the national civil rights movement.

A march was organised from Selma to Montgomery, the state capital, to petition the avowedly racist governor George Wallace. The peaceful marchers were met by state troopers who beat them and rained canisters of tear gas onto the heads of the crowd.[90] Two weeks later eight thousand people assembled at Brown Chapel in Selma and, led by Martin Luther King Jr., they set out again for Montgomery. By the time they arrived on the steps of the State Capitol building four days later,[91] the crowd had swelled to twenty-five thousand.[92] Among their number were Roger Willcox and his twelve-year-old son, Pete.

Dr King addressed the crowd:

> *We must come to see that the end we seek is a society at peace with itself, a society that can live with its conscience. That will be a day not of the white man, not of the black man. That will be the day of man as man. I know you are asking today, 'How long will it take?' I come to say to you this afternoon, however difficult the moment, however frustrating the hour, it will not be long, because truth pressed to earth will rise again. How long? Not long, because no lie can live for ever. How long? Not long, because you still reap what you sow. How long? Not long. Because the arc of the moral universe is long but it bends toward justice.*[93]

Pete later told the journalist Josh Eells that the march instilled in him 'the notion that if you dedicate yourself to something outside your immediate sphere, it's going to be a more fulfilling life'.[94]

Most American schoolchildren are sent to summer camp. When he was fourteen years old the parents of Pete Willcox sent him to the Soviet Union. It was 1967, the year before Pavel Litvinov sat down in Red Square to protest the invasion of Czechoslovakia. For Roger and Elsie the USSR didn't hold the same demonic connotations that it did for most Americans. Pete spent several weeks touring the country, visiting a Russian school and eating austere Russian meals. He took a train from Moscow to Crimea, where he spent six weeks at Camp Artek, the famous retreat for the children of party members and international delegations (Yuri Gagarin, the first man in space, was at the camp that summer[95]). Again and again Pete met Russians who yearned for peace, for better relations with America.

When he was nineteen years old he was drafted to fight in Vietnam. Soon afterwards he took a call from the wife of the folk singer Pete Seeger, a family friend. The Seegers had an offer to make. They owned a boat and used it to teach local children about pollution in the Hudson river. The first mate had recently been tried for draft evasion but had somehow persuaded the judge that the crew of the boat should be exempted from the military. If Pete wanted, he could join the crew and avoid Vietnam.[96]

Pete Willcox eventually became captain of Seeger's boat. He dated the cook Maggy, it was love but it didn't last. She married another guy, but they stayed in touch.

After six years of educating kids about one dirty river in one state, Pete was getting itchy feet. In 1981, at the age of twenty-eight, he answered an advert in a magazine for a job as a deckhand for Greenpeace. On day one, because he was a good painter, he was made the new first mate. Three months later he was captain. Six months after that he was sued for piracy for the first time after taking part in a direct action protest against toxic waste dumping off the New Jersey coast. Two years later came a second piracy charge, this time for boarding a Japanese whaling ship.[97]

Then in 1985, the French secret service bombed his ship, the *Rainbow Warrior*, and killed his friend, the photographer Fernando Pereira.

Two of the bombers – Captain Dominique Prieur and Commander Alain Mafart – were arrested by local police.[98] Three other French agents were arrested by the Australian authorities but were released and picked up by a French submarine.[99] Four months later Prieur and Mafart were sentenced to ten years' imprisonment for manslaughter,[100] but the French government threatened economic sanctions against New Zealand unless the agents were freed.[101] A political deal was struck that saw the bombers held on a French Polynesian island, but after just two years they were released and returned to France, where Mafart was promoted to the rank of lieutenant colonel.[102] It later transpired that President François Mitterrand knew about the plan to bomb the *Rainbow Warrior* before it was executed.[103] The leader of the team that planted the bombs was reported to be Louis-Pierre Dillais. He later moved to the United States and settled in Virginia, where he still lives, running the American arm of a Belgian weapons manufacturer.[104]

Pete was devastated by Fernando's death. He took a year out, sailed a bit, took time to reflect. Eventually he was ready to get back to campaigning. Soon he was being arrested in Denmark, then the Philippines. A movie came out, *The Rainbow Warrior*, in which Pete was played by Oscar-winner Jon Voight. Years passed. Greenpeace grew in size and scope. The job of captain changed.

'When I started out, there was one campaigner on the boat, and half the time his job was to bring the recreational pharmaceuticals,' Pete told Josh Eells from the magazine *Men's Journal*. 'Now there's a campaigner, an assistant campaigner, a comms person, a second comms person, a webbie, photographer … I remember saying in the early eighties that we had to get as disciplined and organised as IBM or Exxon, or we weren't going to matter. And in a lot of ways we did, and it sucks. I'm still glad we did it. But now you're kind of just a cog in a wheel.'[105]

Pete met an Argentine doctor at sea, married her and had two kids, but the marriage didn't last. Later he connected with his birth mother, then in 2010 he was visiting Maine to pick up a classic yacht for delivery when he met up with Maggy, the cook back on Pete Seeger's boat forty years earlier. She was single now, so was he, and that thing between them was still there. Three years later they married. Seven months after that he sailed for the Arctic.

Pete Willcox's diary

1st November
November! Made it! Good things are going to happen this month, but there are still alligators in the pool. So do not count your chickens yet! Got off to a slightly better exercise period. Two sides of the box were walkable. Then the day got a little weird. I got taken to a room up on the fourth floor which was some kind of meeting room. Two guys, one in a suit, one a plain shirt. They said they were from special services or something like that. The suit started to talk but the other guy was such a shitty translator, it was really hard to figure out what they were trying to say. They made the point that I was responsible for the crew being

here. I did not like it, but agreed anyway. They then spent a long time saying if I made a statement things would be better, and we would go home. They asked me what I thought, and I said if that's what they told me I would have to accept it. They did not whip out the paper and start questioning me. The interpreter was not up to it. But at the end I just agreed with them, saying I did not know anything that was going on, and if they said I would get out after making a statement, I would have to believe them. In the end they might have thought they convinced me. But if they think I am making a statement without my lawyer, they are smoking crack.

TWENTY-FOUR

Frank Hewetson's diary

1st November

Just had a 20 minute phone call with Joe + Nina. It was really really lovely. So good to talk to them and hopefully reassured Joe of my well being. Nina pretty damn upset with Greenpeace and how they got it 'so' wrong. I tried to tell Nina that I always knew a prison sentence was coming my way but she is still very angry with them. She has done loads of really good interviews as well. Major radio + news etc. So proud of her. So proud. Things could get a bit ugly if I get a longer jail term here. I think Nina will start demanding a new lawyer + possibly criticise Greenpeace, which would be disastrous on an open forum. God I hope it's over soon for that sake alone.

2nd November

Really strong dream last night. Boris was obviously making noise and rattling the steel bunk with the light on. I was confused and convinced it was Nina coming into our bedroom to go to sleep. Assuredly to do with the phone call and the sudden contact made with back home. It was really upsetting though, because for one second I felt so comfy and back at home in bed. Reality was a hard bite.

I keep thrashing Yuri at chess. It's getting a bit embarrassing.

4th November
Boris' snoring is getting really bad. Started dreaming of toe clip electrodes that would be linked to a decibel meter and apply an equal and appropriate level of electric shock compared with the sound level and resonance of the snoring.

The Investigative Committee is trying to split the Arctic 30.

If they can get some of the activists to give evidence saying who took an active role in the protest at the rig, they can focus the charges on just a few of them. It's a strategy that could see some of them go free but ensure the rest are jailed for many years. And if the IC can get the thirty to turn on each other they'll secure a propaganda triumph for the Kremlin, especially if one of the turncoats is from the famous Litvinov dynasty.

Three days after being freed from the punishment cell, Dima is shaken awake by a guard.

'You, get up, you have a meeting.'

'I do?'

'Your lawyer's here.'

Dima jumps to the ground, then he's marched down the hall with the guard following just behind him. 'Right left right left right left … ' They turn a corner and there, coming towards him, are the two guys from the FSB. *The competent authorities*. The fist clenches in Dima's stomach. Gerbil breaks into a grin.

'Hey, that's Litvinov isn't it?'

'It is, it is,' says his friend with the helmet haircut. 'It's our old friend Litvinov.'

'Ah, but where's he going? That's the question.'

'He's going to see his lawyer.'

'Oh *is* he?'

'He is, he is. He's got a meeting.'

'Ooooh, a meeting. Sounds important.'

'But I think we should have a little chat with him first, don't you?'

'Yes, yes. I think we should. I think the lawyer's going to have to wait a while.'

The guard tugs the back of Dima's shirt, pulling him to a halt. Helmet-hair opens a door and holds out an arm. Dima looks at each of the men in turn then nods and steps through the open door. He's in a small meeting room with a row of chairs facing a single seat across a table, on which sits a bowl of biscuits. It's like the scene of a job interview with five people on the panel. The competent authorities follow him in and shut the door.

Helmet-hair sits down and invites Dima to take the seat opposite him. Gerbil doesn't take a seat, instead he lowers his tiny, bony little arse onto the edge of the table then shuffles along, making himself more comfortable before leaning down and exploding in Dima's face.

'You fucking bitch! You know how many years you're going to be here? You think you're going to get out after two months? Oh my *God*, you're going to be here for years! You understand that, right? Did you like the punishment cell? Oh, you didn't? What, you thought because you're a Litvinov that we can't just make you disappear? Well you know what, you're going to be spending a lot of time in that cell, my friend. What have you got to say to that?'

Dima flinches in the face of Gerbil's onslaught, the knot in his stomach is as tight as it's ever been, it's generating a powerful anxious energy that's running down his legs and up his spine, sucking the breath from his lungs and making his hands shake.

Helmet-hair picks a biscuit from the bowl and slides it between his thick lips, his eyes resting grimly on Dima's face. Silence. Crumbs are dropping from his mouth and falling into the creases in his shirt. He swallows and reaches for another biscuit, bites a chunk out of it and with his mouth full he says, 'You're in our hands now, arsehole. Oh, you thought you had to commit a crime to go to the *kartser*? Buddy, you'll be going to that place any time we want you there. So look, bitch, now's when you start telling us what happened at that platform. No protocol, just you, us and the truth. And you'd better start telling us fucking soon.'

Dima coughs into his hand and takes a breath. 'I'm under strictest orders from my lawyer and—'

Gerbil screws up his face and slaps the air in front of him. 'I'm just so disgusted by this man, I've got no time for this bullshit. He had his chance but he just sealed his fate.' He stands up and opens the door. The guard is still standing outside. 'Get him out of here.'

Helmet-hair pushes the rest of the biscuit between his lips, slaps his hands together, stands up and walks out. Dima gulps. He stares at the empty chairs in front of him then looks up at the guard, who gives a long low whistle.

'Boy, they don't like you, do they?'

Dima is taken back to his cell. He lies on his bunk, staring at the ceiling, going over what just happened. He's still shaken. The fist is like a rock in his belly. They're out to get me, he thinks. There's no doubt about it, not now. This is what they did to Dad.

Nearly half a century has passed since Pavel Litvinov sat down in Red Square in the certain knowledge that he would be punished with the full force of the Soviet state. But right now Dima's thinking of an episode that happened a year before that. The time when his father wrote a letter to the editor of *Izvestia,* the leading Soviet state-controlled newspaper.

I consider it my duty to bring the following to the notice of public opinion. On 26 September 1967 I was summoned to the Committee of State Security to be interviewed by an official of the Committee named Gostev. During our talk another KGB official was present but did not give his name. Immediately the conversation was over, I wrote it down from memory, because I was convinced that it graphically revealed tendencies which should be given publicity and which cannot but cause alarm to progressive public opinion both in our country and throughout the world ...
I protest against behaviour of this sort on the part of the state security organs, behaviour which amounts to unconcealed blackmail. I ask you to publish this letter, so that in case I am arrested, public opinion will be informed about the circumstances leading up to this event.

In the letter Pavel gave his verbatim account of the interrogation, explaining how the KGB agent Gostev warned him not to report details of a recent dissident trial; how the officer accused the dissident of 'hooliganism' despite the fact that the man had merely read out a poem in Mayakovsky Square; how Pavel would himself face trial unless he stopped his political activism.

'Pavel Mikhailovich, we don't intend to have a discussion with you,' Gostev had said. 'We are simply warning you. Just imagine if the whole world were to learn that the grandson of the great diplomat Litvinov is engaged in conduct of this sort. Why, it would be a blot on his memory.'

'Well I don't think he would be against me,' said Pavel. 'May I go?'[106]

Izvestia refused to publish the letter, but it did appear in the *International Herald Tribune*. A year later Pavel was arrested, put on trial and sentenced to exile. And now, half a century on, 4,000 miles from Dima's cell in Murmansk, the retired physics teacher is schooling himself in the system that once targeted him, the same system (for he believes nothing much has changed) that has captured his son.

'It is the Soviet legal machine, so to speak,' says Pavel. 'So I started reading the Russian criminal code – for many years I didn't do it – and the constitution and how it's formulated and how to fight it, and I talked to lawyers and so on. It was important, because in spite of Russia being a totalitarian state they still have to make something by law, and we had to answer every legal step. Some people didn't understand it, but I knew. Because they once did it to me.'

Izvestia, the newspaper he wrote to in 1967 seeking justice, is still in print and is playing a leading role in the propaganda war against his son – even speculating that Greenpeace volunteers may have been responsible for beating up a Russian diplomat in the Netherlands. But *Izvestia* is no longer owned by the Communist Party. Now it's owned by Gazprom.[107]

Dima is standing in the doorway of his cell watching the guards lifting mattresses and looking under them, flicking through the pages of his books and peering inside jars. Then from behind him he hears a familiar voice.

'Ah yes, Litvinov. Dimitri, how are you doing, my friend?'

Dima spins around. It's Popov. He's standing on his toes and looking into Dima's cell.

'They found anything?'

'Nope.'

Popov nods, then almost absently he says, 'So, you'll be leaving us soon.' He drops onto his heels and flashes a golden toothy smile.

Dima's heart jumps. 'We're getting out?'

Popov sniffs. 'No, no. They're moving you.'

'Moving us? Where?'

'St Petersburg.'

'*St Petersburg?*'

'Yup.'

'When?'

'Couple of days maybe. For all I know tomorrow.'

'Do you know why?'

'Dimitri, please.' Popov holds out his hands in a plea of false modesty. 'They tell me nothing. You probably know more than I do.'

'I don't know anything.'

'I'm sorry, I shouldn't have said anything. Silly me.' The governor looks down at his shoes and chews his lip. 'No, it's all a mystery to me. If I was to guess I'd say you're going to be the subject of a bit of special treatment. But then, I know nothing.'

The fist in Dima's stomach clenches. 'What does that mean, special treatment?'

'It means just that. Special treatment.'

'And ... what, it's just me moving? The others, are they coming too? To St Petersburg?'

'I would imagine so, yes.'

'But you don't know for sure?'

Popov shrugs, pats Dima on the shoulder and waltzes off down the hallway, running a finger along the wall as he goes.

As soon as the lights are killed and the *doroga* is up and running, Dima gets the news out. Phil's cellmate pulls in the sock and hands him a note. (See opposite.)

The road buzzes with the news, it's shouted over the walls at the *gulyat*, it's all anyone talks about.

'It's today.'

'I heard it was next week.'

'No, it's definitely tomorrow.'

'Well, you should know this,' Roman shouts to Dima over the wall. 'The *Stolypin*, it is hell.' He uses the Russian slang word for a prison transport train, named after the tsarist prime minister Pyotr Stolypin, who commissioned railway carriages to transport revolutionaries to Siberia. 'They are very fucking horrible. They are bad. They are *very* bad. It can take a month. You stay in transit prisons on the way. Tough places. We will be in a car cell with many people. Not nice people. There's no water, it's very hot, there's no food. We'll

Hey man!
News.
I'm being moved.
to a different jail
in St. Petersburg
Sunday or tomorrow.
My lawyer was
not told. Possib-
ly others also.

DMM

need to bring plastic bottles with us to piss. It is going to be horrible. I've heard about these transports.'

And Dima shouts back, 'Come on, man. Don't worry, we're not even sure we're going by *Stolypin*. Maybe we'll go by bus or by plane.'

'No, it is going to be very bad. I am not happy about this. It is bad. And in St Petersburg they have Kresty' – the notorious prison, famous for housing political detainees as far back as Trotsky – 'It is a tough prison. It has bad cells, poor conditions, a very strict regime. My cellmates, they say it is dark.'

Daniel Simons is terrified.

If he screws this up then he'll be responsible for the Arctic 30 staying in jail for years. He thinks this is the only good shot Greenpeace has at getting everyone out. There is nothing he has dreaded more than appearing as a witness before the International Tribunal for the Law of the Sea.

Twenty-one eminent judges from around the world are sitting in a semi-circle in a vast Hamburg courtroom. All but one of them are elderly men. They look like a council of Jedi Knights in a *Star Wars* movie. And Simons is standing before them.

He and his boss Jasper Teulings have pushed for weeks for a hearing at ITLOS – the international court empowered to demand the release of the *Sunrise* and its crew. Now that hearing is starting and Simons is giving evidence. For the Greenpeace lawyers it looks like an open and shut case – the raid on the *Sunrise* would only have been legal if the Arctic 30 were genuine pirates, and by now not even the Kremlin's spokesmen are claiming that.

The Russian government is boycotting the hearing, a move taken by observers to be an admission of the weakness of their case. Nevertheless one of the judges, Vladimir Golitsyn, is from Russia, and Simons feels he seems to be speaking for the Kremlin as he fires off a volley of hostile questions.[108] Simons reels off answers with a confidence that belies the heavy responsibility he's feeling.

When the submissions are completed the court announces that it will rule in a fortnight. Daniel Simons breathes a sigh of great relief. He leaves the courtroom and fetches his computer. He and Teulings have one more thing to do today.

They're still in the court building. They go to the Dutch delegation and tell them they have something that might be of interest. Simons opens his laptop and presses play. On the screen masked armed commandos abseil onto the deck of the *Arctic Sunrise*. Activists thrust their arms in the air and offer no resistance. Frank is chased up steps and pulled to the ground.

It's the film from Phil, recorded on the camera card and smuggled out through a matchbox.

The Dutch delegation stares open-mouthed at the screen. They're impressed by the intensity of it. The activists look nothing like pirates. Immediately the delegation submits the footage to the international court. Greenpeace submits it to BuzzFeed, and within an hour it's running on TV stations across the globe.

Simons and Teulings are certain the ITLOS panel will see the footage. Nobody's immune to the impact of those images, they think. Not even these unimpeachable judges. It could play a crucial role. That night they toast Phil Ball.

The next morning, 1,700 miles away, Phil is at *gulyat*, walking in circles in a box, when he hears Frank's voice.

'Phil! Hey, Phil, you there?'

'Yeah, I'm here. Still fucking here.'

'You see the TV last night?'

'No, what?'

'That footage you shot. It was all over Russian TV.'

'What footage?'

'Of the raid on the ship.'

Phil stops walking. 'Seriously?'

'All over the news.'

Elation surges through his body. He feels it washing over him, endorphins exploding in his brain. He jumps up and slaps the wall. 'Yeeees!'

'Didn't you see it?'

'My cellmate doesn't let me watch the news. He's into these crappy soap operas. Jesus, I missed the world premiere of my own fucking film.'

'It was amazing. Soldiers coming down the rope, guns, me getting roughed up. I'm on TV getting pushed over. Seriously, they look like thugs.'

'Amazing.'

'Phil, how did you do it? Last I heard, you'd shoved it in the extractor fan in the galley.'

Phil looks up. Through the wire mesh he can see the guard patrolling on the bridge above. Around him he can hear shouted conversations in Russian.

'Come on, Phil. How did you do it?'

'You know what, Frank. When we get out of this place you can buy me a beer and I'll tell you.'

Then, before the day is out, Putin's own Presidential Council for Civil Society and Human Rights, led by Mikhail Fedotov, announces it has written to the head of the Investigative Committee offering to act as guarantors for the Arctic 30 if they are released on bail. It's the same offer that Kumi Naidoo made, but this time it's coming from someone inside the Kremlin.

Pete Willcox's diary

6th November

The IMO Tribunal [International Tribunal for the Law of the Sea] meets today. They are supposed to have a decision by the 21st. That's two weeks. I asked if there was any chance the court would hold us over for another 60 days … and [the lawyer] Alexandre de Moscow said he EXPECTED IT. Boy he is changing his tune! He now says Russia will want to think about the ruling for at least a month. So there goes Thanksgiving, and probably Christmas. And I went right down the tubes. Sounds like we will catch a two day train to St Pete on Saturday. If we could all be together in one car on the train, it would be wonderful. But it does not seem likely at the moment. Came back with JB [New Zealander Jon Beauchamp] & Marco Polo [Kruso]. I got left in a holding cell again for an hour. At least I was alone. I wonder what I did to piss off the front gate guards. We were on the news tonight. I wonder what it was?

The Kremlin appears contemptuous of the ITLOS legal case. Even if it rules against Russia there's no guarantee the order to release them won't just be ignored. And anyway, the Arctic 30 are far more concerned by the imminent move to St Petersburg. They've been told by their lawyers that they'll be split up. Most of the men will be held in Kresty, some of them will go to another prison and the women will be held at the all-female St Petersburg SIZO-5. And still nobody can tell them when the move will happen, how they'll be transported and, most importantly, why. Even the guards ask the activists the same questions. Of course everybody has a theory, especially their Russian cellmates.

'They're getting ready to release you, but they want to sweeten you up first, get you out of this shithole, make sure none of you get beaten up, no black eyes for the cameras.'

'I guess they're getting ready to really screw you, send you down for years, put you in a real prison.'

'It's because the trial is going to start. They're going to have a trial and they want it in a big city, big lights, big show.'

The activists join in the speculation – nothing can stop them doing that – but they also know not to believe anything until it happens.

Don't Trust Don't Fear Don't Beg.

The women are worried that contact with the others will be restricted in St Petersburg. At SIZO-1 during the daily walks they're able to communicate with their friends. It's the best part of the day, shouting over the walls in those dark and cramped boxes and being able to hear the voices of the others coming back. But what if they can't talk to each other in St Petersburg? What if they're alone?

Pete Willcox's diary

9th November

Quiet day. Exercise pit 7. Then around 4.30pm, the jail rights lady came for a visit. She had a one page typed paper on how to survive the move to St Pete. She said 90 or 95% chance we will go in plane or passenger train. But if we go by prisoner train, then we are fucked. The prisoner train takes 2 to 3 days. There are cells with four bunks in them, but they often put 20 prisoners in them. And they don't let you out to go to the bathroom. I am praying ...

TWENTY-FIVE

Frank Hewetson's diary

10th November

Just got confirmed by guard + doctor that transport will be all 30 of us and tomorrow. I can tell Boris is a bit upset with my departure. I guess I must have brought a distinct change and foreign flavour to the cell for the last 42 days. He enjoyed it I'm sure. For Boris time just drifts by, more or less like his life will.

What awaits us in St P and how long we'll be there is the next venture.

Frank is emptying hoarded Valium tablets into the palm of his hand, wondering how the hell he'll smuggle them into Kresty. The guards have just come round saying the move is happening tonight. He has to pack.

After thinking for a while he starts feeding the pills into the lining of his jacket.

'No, no, no,' says Yuri. 'Bad, Frank, bad. They search. You go to *kartser* in Kresty. No no, put here.'

And Yuri points at a little flap in Frank's trousers – a tiny pocket inside another pocket just above his knee.

'Here. Better.'

Across the prison the activists are packing their bags, saying goodbye to their cellmates and writing final letters to friends and family, unsure if they'll be able to communicate from St Petersburg, hoping Mr Babinski can get their messages to their loved ones despite Popov's crackdown.

It's 4 a.m. The guard is waiting at the open door. Time for Frank to say goodbye. He turns around and shakes his cellmates' hands.

'Boris, Yuri, thank you.'

'Goodbye my friend Frank.'

'Good luck in Kresty. We like very much having you here.'

Frank pumps their hands then he thinks, no, this isn't enough. He goes to hug Boris but the Russian jumps back with a look of panic on his face.

'Oh, stop being so damn homophobic!' Frank throws his arms around Boris. Slowly the Russian raises his hands and pats Frank on the back. 'Thank you,' says Frank. 'Thank you for welcoming me to your home. I'll never forget how you treated me here. Never.'

Then Frank hugs Yuri. 'Thank you for what you did for me. You taught me so much. I was very lucky to be put in with you.'

'Thank you my friend,' says Yuri, whose face is buried in Frank's shoulder. 'We will miss you.'

The activists are taken out floor by floor. For the last time they cross the yard. The lights of SIZO-1 are blazing around them. The *doroga* is cooking, the prison is living, it's the same scene they faced when they arrived here seven weeks ago. Dima stops and turns around and faces the windows. He lifts a fist into the air and shouts a single word.

'AUE!'

AUE. Pronounced *ah-oo-yeah*. It's an abbreviation, it stands for *Arestanskoe Urkaganskoe Edinstvo*. Translation: Arrestees Criminal Union, the society of the incarcerated, a banned exclamation to identify yourself with the thieves in their struggle against the stars. A call to non-submission.

'AUE! Vitaly, AUE! My cellmate, farewell, may you see freedom soon!'

And from the windows comes a wall of noise.

'AUE!'

'Go pirates!'

'Your freedom must come!'

'AUE!'

The Arctic 30 are leaving SIZO-1 with the solidarity of the prisoner community ringing in their ears. In single file the crew walk towards a waiting *avtozak*, their bags slung over their shoulders, their breath lit by spotlights as it freezes on the cold night air. They glance back to take a final look at Murmansk isolation prison.

Some of the crew take a moment to think about the cellmates they're leaving behind, men who in many ways they've come to like and respect. Ivan, Boris, Yuri, Vitaly. Some of them are imprisoned because they've done bad things to good people, but in SIZO-1 they did what they could to help strangers survive.

The activists are slipping from Popov's grasp, but what awaits them in St Petersburg? Can Kresty be as bad as their cellmates claim? And are they being taken to a place that will be their home for the next seven years of their lives?

Next to the bus stands an officer in a sharp camou-flage uniform. 'I am in charge until we hand you over in St Petersburg,' he announces. 'There are rules and I expect you to obey them. My men expect nothing less than your absolute co-operation. Our journey is by *Stolypin*. I cannot tell you how long it will take. As long as you recognise our absolute authority, you will be treated well. I see no reason why this should be unpleasant for any of us. Okay, let's go.'

They're loaded onto the bus, it pulls away and they drive through the gates. Behind them the shouts and screams from the windows die out and all they can hear is the crunch of tyres on grit as they leave SIZO-1.

The *Stolypin* is comprised of old-fashioned carriages hooked onto the back of a passenger train. The corridor goes down one side of the carriage, with cells along the other. There are no windows in the cells, just shelves, like benches, two rows of four in each. Only the top ones have enough headroom. The windows on the corridor are frosted glass with little gaps that the activists can see through. At one point they glimpse birch trees with snow on the ground, then they pass some buildings and they see the lights of a town.

It's like *Doctor Zhivago*, Frank thinks. Everyone banged up on the train, and this carriage is pretty much from the same era. Maybe Julie Christie's going to turn up.

They can touch each other, they can speak to each other without shouting, without a guard interrupting. For some of them, it's a revelation.

The forty-year-old Argentine sailor Hernan Orsi pulls out a sheet of paper and a pen. 'Okay,' he shouts out, 'here's your chance to say who you want to play you in the movie of this shit. Me, I'm getting played by Benicio Del Toro. Colin, you're Jeremy Irons.'

'Jeremy fucking Irons? Are you kidding me?'

'Andrey, you're Gérard Depardieu.'

'No no,' someone crises out. 'Andrey is Dustin Hoffman.'

'Okay, Dustin Hoffman.'

It takes an hour for the train to agree on a full cast list – Frank is Jason Statham, Alex is Jennifer Connelly, Camila is Jennifer Garner, Sini is Naomi Watts, Phil is Jude Law, Kieron is Orlando Bloom and Dima is Jean Reno, the kindly assassin from the movie *Léon*. Pete is Jon Voight – 'What? Again?' – and Denis is 'Sickboy in *Trainspotting*' (aka Jonny Lee Miller, who was once married to Voight's daughter Angelina Jolie, making Pete a sort of celebrity father-in-law to Denis).

The women are in cages together. They play games all day, they don't want to sleep because they don't know when they'll be this close again. They hold hands and talk – about their lives back home, boyfriends and families. Camila has photographs with her, of her parents and her brothers and sisters. She shows them to the others, then she reads out a letter her father sent her, translating the words as she goes.

'Dear Bochi,' she says. 'That's what he calls me. Bochi. It means bold. He says I was bold when I was born. Okay, so … Dear Bochi, The things you are fighting for are worth the risk. You cannot imagine how proud your mother and I are, seeing how you've grown up with the values, the solidarity and the humility we tried to raise you with. When people ask me how I'm doing, I tell them I've seen the most beautiful flower in the world, and she is mine.'

Alex and Sini stare at Camila then throw their arms around her.

In Amsterdam Faiza's mother Mimount is trying to process the move, fearing the worst. 'I wondered why they were being transferred to St Petersburg. I was sceptical, I was doubtful of the Russians' intentions, and there was absolutely no information on why this was happening. I was scared it might be bad news. I was afraid, afraid that they were being taken to a permanent detention centre. I feared for my daughter. I wanted to take her place because I feared for her.'

It's Anthony's birthday, and for a present Frank has given him a Valium tablet. It takes the top of his head off. He sleeps for sixteen hours straight. When he wakes up he tells Frank it was one of the best birthday presents he's ever had.

This *Stolypin* has a toilet, and visiting it is a joy for some of them. They get to see all of their friends, the people they never spoke to in prison. As they walk down the long thin corridor in front of a guard, hands poke through the bars, so they touch them as they walk, like a rock star running down the front row at a concert.

But for some of the thirty, the journey is a chance to air frustration, even anger, at Greenpeace. Some of the activists think they're only in jail because of a monumental internal mistake. They're hurting and they have something to say. It's the first time they've all been in such close proximity since coming off the *Sunrise* and there's tension between Frank and two others. They're pointing the finger at him and Dima.

Frank was the co-ordinator of the protest, Dima was the lead campaigner on the ship. Ultimately the action was their responsibility. Now Frank is in a compartment with two guys who want to know why the piracy charge wasn't predicted. The argument goes back and forth. Nobody's shouting, but this is heavy.

'You dumped us in this shit, Frank. You need to face up to what you did. You were responsible for the action, you were in charge, this is your fault. You and Dima, you're to blame. Bringing a bunch of activists to Russia, messing with the FSB. What did you think was gonna happen?'

'What do you mean I brought us to Russia? We weren't in Russia, it was international waters. Everyone knew the score.'

'Nobody knew. That's the point, isn't it? Nobody knew because nobody stopped to think how heavy it could get. And that was your job.'

It's a line that's been running in the global media – that Greenpeace should have known an action on a Russian oil platform would be met with a legal hammer. Four weeks ago Dominic Lawson – the son of Lord Nigel Lawson, the UK's most prominent climate change sceptic – took a full page in the *Daily Mail*, the world's most read newspaper, to eviscerate the organisation.[109]

I wonder what they had imagined would be the reaction of the Russian coastguard, defending the security of what in any waters would be a highly sensitive installation. They had been warned to back off in the most explicit terms and told that any attempt to scale the exploration rig would be regarded as a hostile act … Faced with the unenviable imaginary choice of a government run by Vladimir Putin and one run by Greenpeace, I would vote for the former every time. Putin might be a vengeful and autocratic ruler

*in the Russian tradition; but he is not part of a gang of
well-meaning fools seeking to drive mankind back into
pre-industrial poverty.*

In a film for the hugely influential *Sunday Times* newspaper
entitled 'Should we blame Greenpeace?'[110] the commentator
Rod Liddle said:

> *So Greenpeace authorised their people to clamber aboard
> a Russian gas installation which the Russians consider a
> security installation, and Greenpeace are surprised the
> Russians didn't accept all of this in high spirits. What on
> earth was Greenpeace thinking?*

Liddle then interviewed the former Russian presidential
adviser Alexander Nekrassov.

'Do you have any sympathy for the Greenpeace people?'
he asked.

'I think the Greenpeace people made one crucial mistake,'
Nekrassov replied. 'They should have realised the gravity of
the situation. They should have warned all the people involved
that this is not an easy thing, it carries a lot of danger here.
I feel sorry for the people who went there without knowing
what they were in for.'

Similar stories have appeared in newspapers and on
television around the world. It's a line that Dima heard
on state-run TV in his cell at SIZO-1. And he's picked up
some hostility from a few of the other activists, a sense that
because he's Russian he should have known how the regime
would react.

But from where Dima is sitting, this is the third time he's been arrested in the same waters, going back to 1990. A year before the *Sunrise* was stormed, he was one of the people who protested at the same rig. Putin was president back then and they weren't touched by the authorities. Dima wasn't alone thinking Russia had a fairly sane and rational government. Granted, there were voices saying Putin was mad, but they were far from the mainstream, closer to a lunatic fringe. At worst Dima expected the crew would be detained for a few days aboard the ship after the action. And even in jail he stands by that expectation, given what he knew then.

In Frank's compartment the argument is still raging.

'Come on, Frank. When we had a legal briefing about what the charges could be, I'm not sure piracy was even mentioned.'

'It fucking was.'

'Well, even if it was, it was never said to us like it was a serious risk.'

'Because it wasn't. Because we weren't going to commit piracy. You can't commit piracy on that platform, it's attached to the seabed, it's not a ship. Jesus, guys, I know we're in the shit here. I know we're all hurting, but please, give me a break with this blame game.'

'Well, if it wasn't a serious risk then what are we all doing here?'

The row peters out to silence, but the tension hangs unresolved as the train rattles through the night.

Alex's cousin has sent her a Russian–English translation booklet and it has a section devoted to romance. She pulls it out on the prison train. It has the Russian translation for phrases including *I would die without your love* and *I like it when you touch me here* and *I want you.* Alex, Sini and Camila shout out these lines at the guards then collapse in laughter. But the guards don't smile. They stare ahead with concrete expressions.

Twenty-four hours have passed. The head of the guards tells them they're close now. For some it's a moment of huge relief. So they're not being taken to a transit prison. This isn't the *Stolypin* journey of their nightmares. Then the realisation dawns that they're hours away from being split up. Alex and Sini shout out, 'We love you' and, 'Wherever you go, stay strong!'

The train pulls into St Petersburg and the carriages fall silent. The guards take them out, one cell at a time. They have to jump across the track from the train to the open doors of an *avtozak*. Sini and Alex are pushed into a compartment together. Alex zips up her purple ski jacket and reaches out for Sini's hand. They squeeze as tightly as they can.

The men are loaded into different vans. The drivers turn the engines. The women are heading for SIZO-5, some of the men for SIZO-4. But most of them are driven to St Petersburg SIZO-1.

Kresty.

It's a crumbling stack of red brick and barred windows on the banks of the river Neva and would look like a derelict Victorian mill or an asylum if not for the distinctive onion domes of the Alexander Nevsky cathedral that rise from the middle of the complex. Kresty means 'crosses'. The nine-teenth-century architect who designed the prison had it built in the shape of two crucifixes, so the prisoners could better repent of their sins.[III]

The prison housed the enemies of tsarism, then after Lenin's seizure of power Kresty housed enemies of the new communist government. In *The Gulag Archipelago*, Solzhenitsyn wrote: 'During the whole year [1918] it would certainly seem that more than a thousand were shot in Kresty alone' – of whom six were peasants guilty of clipping excess hay from a communal farm to feed their cows.

During the long decades of Soviet rule Kresty held both criminals and political dissidents from what was then called Leningrad, Russia's most liberal city. Anna Akhmatova, one of Russia's greatest poets, would wait for hours outside the prison walls in the hope of delivering a package to her son, the historian Lev Gumilev.[112] Akhmatova captured the anguish of the relatives of Kresty's inmates in her famous poem, 'Requiem'.

How, the three hundredth in a queue,
You'd stand at the prison gate
And with your hot tears
Burn through the New-Year ice.

In the preface to the poem, she describes how she was prompted to put her experience into words after spending seventeen months outside the prison gates. She saw 'a woman with blue lips. She had, of course, never heard of me; but she suddenly came out of that trance so common to us all and whispered in my ear (everybody spoke in whispers there): "Can you describe this?" And I said, "Yes, I can."'

Escape from Kresty is almost impossible, but many have tried. In 1946 a prisoner named Volkov removed bricks from the wall of his cell, one by one, and put them in his chamber pot, which he emptied outside. Eventually he created a hole to the street, and one night he made his break. Having secured his freedom he went immediately to the public baths, where he was recognised by a surprised but diligent off-duty guard. It is said that Volkov was returned to Kresty humiliated but clean.[113]

In 1984 two prisoners forged KGB identification papers from cardboard and red thread. They scoured magazines for pictures of uniformed officers who resembled them, cut out those photos and stuck them to the card. Their counterfeiting skills were so advanced that nobody at the gates of Kresty thought twice about letting them stroll out into the street to embrace freedom.[114]

The Soviet Union was dissolved on 26 December 1991, but Kresty remained. Four years later, under the presidency of Boris Yeltsin, a monument was erected on the opposite bank of the River Neva, commemorating the victims of political repression. A year after that another monument was installed in Kresty itself, this time to the poet Anna Akhmatova. But recognition of past crimes did little to improve conditions there. By the end of the millennium Kresty housed ten thousand prisoners – three times its capacity – and was suffering a virulent outbreak of tuberculosis.[115]

As the *avtozak* pulls into the grounds of the famous prison the men see barbed wire and the decrepit, dominating cathedral that rises up from the middle of the jail, built in the same style and with the same bricks as the incarceration blocks. At the top of a white dome are two crosses. One is a typical Russian Orthodox cross, the other a more familiar T shape. From a certain angle the two crosses, overlaid, take on a very particular profile, and Dima remembers that it's this shape that is the design of the prison tattoo that professional criminals have inked around their fingers across Russia – the U of the Russian Orthodox cross, over a T.

Gathered in the reception area are fourteen of the Arctic 30, including Pete, Frank, Phil, Roman and Dima. The rest of the men have been taken to St Petersburg SIZO-4, an isolation unit across town. The eight women are being taken to the all-female SIZO-5.

The activists at Kresty are processed, then led to their new cells. Dima steps through the door and puts down his pink bag. The door slams behind him. The smell of fresh paint fills his nostrils. A man gets up from his bunk and introduces himself as Vasily. He's in his mid-thirties, well-built, very tall, wearing expensive sports gear.

'Welcome.'

'Thank you.'

'Please, sit down.'

'Thanks.

'Would you like a cup of tea?'

'Sure.'

Vasily bends down in the corner of the cell and produces an object that makes Dima visibly flinch. 'Wow, that's a big motherfucker of a tea boiler.'

'Thank you.'

'In Murmansk we had these little sort of coils that you put inside the glass. But that looks like a ten-watt massive fucking tea-boiling pot. Are you even allowed that in here?'

The man shrugs, 'Of course not.'

Then, while the water is boiling, Vasily says, 'Dima, we do not eat prison food here in my cell.'

'No?'

'No, we do not.'

He opens up the fridge – 'We have a *fridge* in our cell?' – and pulls out a French baguette, cuts off a slice, spreads a knob of butter over it, produces a jar of red caviar, spreads a spoonful onto the bread, closes the sandwich with a flourish and holds it out. And Dima thinks, no, we're not in Murmansk any more.

Vasily tells Dima his story. He rented out a house to some students who grew nine marijuana plants, and the investigators say he knew all about it. He's been in Kresty a little over a year, waiting for his trial on narcotics charges. But Dima knows enough about Russia to suspect that's not what Vasily is really in for. The clothes, the food, the fridge, the attitude. This guy's a bandit. Rich. Influential. Sorted. A post-Soviet killer-type bandit.

'The cells were all painted two weeks ago,' says Vasily. 'Refitted entirely for you. Kresty has the *doroga* but you aren't allowed to be part of it. The guards told us, "If you want to stay in these lovely cells with the Greenpeace people, behave."'

Then he shows Dima an order book for the prison shop. It has five pages of food to choose from.

'You order on one day and get it the next day. It works. For people like me anyway. And you too, Dima. You too.'

In a cell down the corridor Frank is sitting on his bunk and looking out through the window. He can see the onion domes of the cathedral through the bars. He imagines Trotsky being marched down the corridors. Many people died here, Frank knows that. Many political prisoners. He can feel the weight of history. The weight of hopelessness. Frank was brought up a Catholic and has a fondness for ornate churches. This cathedral in the middle of the prison looks beautiful, but utterly incongruous in its setting. He writes a note to the governor requesting a tour of the church and hands it to a guard.

Pete Willcox is sitting on the edge of his new bed, writing in his diary.

12th November
Got to the prison around noon, suspected and neglected and put away. The cell is about half the size and I have a roommate. So it sucks. The view though has a bit of a large canal out to the left and a Russian Orthodox church to the right. Reminds me of George's Cross we put up on Amchitka. I am not doing very well. Is this shit ever going to end? My roomy Igor is twice my size but very nice. He has been here 18 months. I do not know what for. He is very hygienic. Tonight he looked at dinner and said it was garp [shit]. So he took the sardines, pulled out the bones, and made a very nice sardine and noodle dinner. We do not talk much. All in all, I would much rather be back in Murmansk.

Frank Hewetson's diary

12th November
This is the oldest and biggest prison in Russia and is more or less a museum. I had a nasty stand-off with the same guy who gave me shit on the train, while waiting

*for allocation of rooms. He blames me and Dima for the
lengthy incarceration. Claims he never got a full legal
briefing and claimed whatever we did turned to shit. He
will send a complaint about it to IMAD [Greenpeace
ships unit]. He made a comment about me never sailing
with Greenpeace again. I think I prompted that by
asking him why he was whingeing so much since he was
the one who swore he'd never sail with us again. Some
people are upset with the way it's turned out and the
campaign team in general.*

*Got packed to my cell and now sharing with Anton.
He got a book out and pointed to the word 'AMNESTIA'
indicating that a good chance for our release would be the
December amnesty, which is a whole month from today,
which I think I can easily handle. BUT with difficulty
if we have continuing rumour see/saw. What we need is
a date to work down to. Just saw long clip on Arctic 30
arrival in St P with the press saying it was like a Top Secret
transport of high level personnel. The circus continues.*

Across town, in a hallway at SIZO-5, the women are lined up
in front of cell doors. The guard standing next to Sini slides
a key into the lock and the door swings open. Sini takes a
nervous step forward. The door closes. Two women look up
from their bunks. Sini tries to smile. She holds out a hand but
the women ignore it.

They're both Russian, middle-aged. They look away. Sini sees
a spare bunk, crosses the cell, sits down and stares at the floor.

Along the hallway Alex is in her new cell with three other
prisoners. One of them is an old woman, tired and sad, short

and plump with long white hair. Alex is sitting on her bed
watching her. A key turns in the lock, the woman jumps up
frantically, makes her bed then stands stiffly in front of it,
waiting for inspection. The guard pokes his head around the
corner then disappears. Twenty minutes later there's a sound
from the corridor, Alex stands up, the woman pushes her out
of the way to get to her bed and straighten the sheet.

The woman sits on her bed. Alex sits down next to her.
She has a dictionary and by pointing at words she can make
herself understood. She asks the woman how she ended up
here, and nervously, by pointing at words herself, the woman
explains that she's seventy-four years old. She says an intruder
broke into her house, he kicked and punched her but she ran
to the kitchen and pulled a knife from a drawer and stabbed
him in the shoulder.

She's looking at seven years. Her name is Marina.

Meanwhile Sini is lying on her bunk, watching her new
cellmates. One of the women is floating around the cell, her
eyes vacant like there's nothing behind them. Sini stands up
and walks to the window. The woman follows her, talking to
herself in staccato Russian. She pokes Sini, shakes her head,
wags a finger. Sini waves her away and sits back on her bed.
But when she stands up again the woman shuffles behind her,
shaking her head with disapproval and muttering to herself
through tight dry lips. Sini unzips her bag and starts laying
her meagre possessions on her bed, the woman rushes across
the cell with a panicked expression, shaking her head wildly.
Sini shoos her away. She lies on her bunk, pulls the sheet
up over her head and holds herself tight. She can sense the
woman standing over her. She can hear her breathing.

She's taken to *gulyat* and shouts over the wall. 'Camila! Alex! Faiza!' But there's no answer, so she shouts again. A guard opens the door of her box.

'I don't know why you're shouting,' he says. 'There's nobody there.'

TWENTY-SIX

Pete Willcox's diary

13th November

In the cell. It sucks. Not freaking but way bummed. Went to the exercise yard. At least you can see some sky, but it is shitty. Saw the isolation cells on the way. They looked really bad. Got a copy of the prison rules in English. Tried to read them but quit when I got to the place where the family of the inmate can recover his body. The prosecutor (new for St Pete) stopped by after dinner. They all had a good laugh when I said Igor 'killed' me at chess today (they saw the board out). I almost asked him what he was going to do with us. But I chickened out. I did not like seeing him.

On the afternoon of the second day in St Petersburg the SIZO governors tour the activists' cells, bringing with them a retinue of officials. When the door of Alex's cell opens, Marina jumps to attention with a look of terror on her face.

Fifteen men file in, one after the other, prison guards and the head of the prison.

'Do you have any problems?' asks the governor.

'No,' says Alex. 'I just want to go home.'

The man sniffs. He looks around. His eyes fall on the old woman. He walks over to Marina and starts asking her questions. She speaks quickly, he nods and makes some notes. The delegation leaves, the cell falls silent. Then Marina starts to weep.

'What's wrong?' asks Alex, putting an arm around her. 'Don't cry Marina. What's wrong?'

One of the other women, a young Russian, sits on the bed and puts a hand on Marina's shoulder. 'She cries because she's been here two years. She asks for help many times and receives nothing. Then they visit you and for first time she speaks to governor. First time he knows who she is. First time they are nice to her. First time.'

In the other jails – Kresty and SIZO-4 – the activists are also being visited by the delegation of officials. An array of huge hats, immaculate uniforms and medals. There's barely enough room for them in the cells.

They ask Dima, 'How is it?' and, 'Is it okay?' and, 'Is it all to your liking?'

'Apart from being in jail, it's all fine.'

The men nod and pat each other on the back then stream out of the cell.

On day three, the activists are visited by the chief prosecutor of the St Petersburg region, accompanied by a colonel and a phalanx of junior officers. The head of the regional human rights commission brings up the rear with the head of the prison. In each cell the chief prosecutor asks the same questions.

'Is everything okay?'

'Do you have any complaints?'

'Are you satisfied with the conditions?'

And Roman tells them, 'Every time, every day, you ask us this. Well, my answer is I don't have any fucking complaints about the conditions, it's not a sanatorium, the only complaint I have is that I'm here at all, together with my twenty-nine friends.'

Frank Hewetson's diary

14th November

Went to sleep early last night + woke up really early thinking every noise outside was the porridge trolley. Feeling the blues today a fair bit. Guess it may be the '3 day' transition period that I feel it takes to get grounded. However I feel down. The isolation is hard. Much more so than Murmansk as conversation over the walls at 'gulyat' is nigh on impossible. Been thinking about Mama a fair bit which always brings me very close to tears. I keep feeling she may die while I'm in prison, which would be unbearable. The cells are more comfortable but the complete lack of contact with the other GP crew is miserable.

Day Four. In each cell the door opens and in come two generals, a lieutenant general and a major general, accompanied by the chief prosecutor, the head of the prison and a clutch of civilians. One by one this collection of sharp suits and enormous hats – this gang that resembles the reviewing party at a Red Square May Day parade – files in and out of the activists' cells.

One of the civilians appears to command the respect of the delegation. The man introduces himself to Dima. 'My name is Fedotov. I am Vladimir Putin's presidential adviser on human rights. My colleague here, the general, is head of the prison system for the Russian Federation, ministerial level. Do you have any complaints? Any questions? How are the conditions here? Oh, I see you have plenty of shelf space.'

'Nice to meet you,' Dima replies. 'Actually, I do have a complaint.'

'Go on.'

'Well, I've been jailed on trumped-up charges.'

'Yes yes, very unfortunate. I understand your concern.'

'The investigators tell me I'll be here for many years.'

'Oh really?' Fedotov glances at the general, his bottom lip protrudes then he looks back at Dima. 'And these investigators, did they tell you what the President says about your case?'

'Actually, no, they didn't.'

'Do you think they've spoken to the President about your case?'

'I suspect not.'

'Well I have.'

'And?'

Fedotov looks around the cell. 'You have a lot of things in here. The conditions here are good.' Then he turns on his heels and walks out, a long trail of uniforms and suits following him, until they have all left and the cell door slams shut. Vasily looks at Dima, his mouth wide open. And he says, 'Who next in our little home? Putin?'

The legal team by now has a new lead lawyer, Andrey Suchkov, a criminal litigator experienced in running successful legal strategies that are inconvenient to powerful interests. His first move on taking over is to establish a line of communication with the lead investigator, and through that channel he soon learns devastating news.

Mads Christensen comes on the video link to address the core team. He looks tired. He lifts his glasses and rubs his eyes.

'Okay everybody, listen up. I've got some bad news. This new lawyer we've got, he's found something out, and … look, it's bad, okay.'

The teams in Moscow, London and Amsterdam share concerned glances.

'This lawyer's been speaking to the investigators,' says Christensen. 'They've told him that when the current detention period expires, they're going to apply to keep them in prison. Another three months. Nobody's getting out.'

Arms wrap around heads. Eyes well up.

'Nooooo.'

'Seriously?'

'Shit shit shit.'

The Kremlin is doubling down. The campaign has failed.

Ben Ayliffe heads up a team that has organised hundreds of protests in dozens of cities. He's run an operation that has energised people across the globe and signed up nearly two million people to the campaign. But now he's slumped in his chair in the Room of Doom, shaking his head. 'What are we actually doing here?' he mumbles. 'Was all this for nothing?'

For families and campaigners around the world the news hits like a hammer. The Arctic 30 are nine days out from the end of the current two-month detention period, the ruling of the ITLOS court is due around the same time, the campaign is a global phenomenon – world leaders, celebrities, newspapers and millions of people are lining up to demand freedom for the thirty. But now Putin has shrugged his shoulders and slammed the door.

In the Russian office the staff have been spied on, lied to, lied about and abused. They have been on the verge of being raided, shut down and arrested. Many of them have come under pressure from their families to quit the campaign, to resign from the organisation. They're branded enemies of Putin now, with all that entails for themselves and their futures. But still they came, every morning, to battle the state media and the FSB and those claims that the Arctic 30 are agents of foreign intelligence agencies determined to undermine Russian economic development. And now they're back where they started. No, worse, the position of the Kremlin has hardened. It feels like it's all been a waste. All of it.

Mads Christensen is back on the video link, speaking to his core team. The mood is dark. He has to tell them something to convince them that the fight is not over, that there are still things they can do. 'We need to increase the strain on the investigators,' he says. 'Something to make them totally fed up with this case. Something that makes them wish we'd just go away. This is a war of attrition and we're more determined. And now we've got nothing left to lose.'

He says the legal team has been waiting for the right moment to apply for bail and land the investigators with a mountain of paperwork and a logistical nightmare, and since the charges were re-qualified there's been a legal justification for doing it. Bail was refused back in Murmansk when they were accused of piracy, but that was two months ago, and now they're charged with a less serious crime.

'Bail is such a rare thing in Russia,' Christensen tells his team. 'It almost never happens. I want to caution you all, this is a long shot. But if we do it we'll cause the Russians a real headache. We'll make them bring everyone to court and defend the arrest and the charges all over again. We need to grind them down. We'll have to offer a bond and it'll need to be a high figure, so the courts can't say we're not serious. But I think we should do it. We're going to apply for bail.'

There is a calendar in the Room of Doom that takes up an entire wall. Key upcoming moments are marked in red pen, and around the third week of November there is a riot of scarlet ink. The two-month detention period – handed down in Murmansk – expires on the twenty-fourth. On the twenty-second the ITLOS international court is set to rule on the Dutch application to have the crew released. In the days before that, the FSB will be applying for that three-month extension of detention. And now the Greenpeace lawyers will be piggybacking onto those hearings to apply for bail. The first case will be heard on 18 November.

Mads Christensen decides to offer two million roubles – fifty thousand euros – for every prisoner. But the figure is academic. Hardly anybody thinks the crew will actually get bail.

The news spreads through the SIZOs. The Investigative Committee is keeping them in jail. Sini is told by the Finnish consul, then she goes back to her cell, lies on her bunk and cries all day, cries until she's so tired she drifts off to sleep with wet and red-raw eyes. She could have done three more months in Murmansk. She could have survived that, with the tapping on the pipe and the shouts over the wall at *gulyat*. But here in St Petersburg, where she has almost no contact with her friends, she's not so sure she'll make it.

Camila hears the news from her lawyer, but by now she feels strong. She feels like she'll get through this, she knows she'll survive. It means she's going to spend Christmas here, and she's going to spend the Argentine summer here. But she accepts that. Okay, shit, so it's going to be three months more, she thinks. Whatever.

Frank is taken to the meeting room at Kresty, where two officials from the British consulate are waiting for him. They have books, copies of British newspapers and his favourite magazine, *Private Eye*. As Frank leafs through the pages of one of the books – it's about English football – one of the consuls shifts in his seat and says awkwardly, 'So they've applied to extend your detention. Another three months.'

Frank looks up. 'What do you mean three months?'

'You're not getting out anytime soon. I'm sorry.'

Frank crumples, his shoulders slump, his chin drops and he stares at the ground between his boots. His mind races for a moment, it's performing calculations, working out all the birthdays that are coming up, the ones he's going to miss. His son's fourteenth, his daughter's seventeenth, his partner's fiftieth, his mother's ninetieth.

He wraps his arms around his head and bites his lip. He fucked up, he knows that now. It was never heroism. Bravado, maybe, but not heroism. He thought he was inside for doing the right thing, but now he feels reckless. Irresponsible. Selfish. All those missed birthdays.

The consul asks him something but he can't speak. He can't even see the man. All he can see are the faces of his kids.

The consul leans forward.

'Frank,' he says. 'You have to listen to me.'

'Sorry?'

'I said, don't you think it's time for the apology?'

Frank blinks. 'Sorry, what?'

'We need to think about an apology.'

'Apology?'

'It's time.'

'Apologise to who?'

'Come on, Frank. You know.'

'Apologise to the FSB? To Putin?' A spark of indignation fires up somewhere near the back of his skull. 'Apologise to Putin? You have to be kidding me.' He sits up straight. 'What we did was right. It was fundamentally right.' He bites his lip and shakes his head. 'No way. I'll do the time. I've got nothing to say sorry for. Nothing.'

TWENTY-SEVEN

Frank Hewetson's diary

18th November

Just had gulyat and brief shout with Denis. He is off to court this morning to receive notification in court of 3 month extension. What appals me is that he is a Russian citizen, living in Moscow. Not exactly a flight risk!

A letter just got delivered for Anton. It was from his lawyer and did not look like good news. He was quiet a long while. After putting it down he washed his face and towelled himself dry. I'm sure he cried. My god I can only feel lucky not being him. He's going to the ZONA [labour camp] and the sentence is likely to be 10 years. He's in his bunk above me, very quiet.

Pete Willcox's diary

18th November

Got up this morning and saw a beautiful full moon over the church. My cell in Murmansk faced north, so I never saw either the sun or the moon. I have now decided to take the heavenly observations of the last two days as a good sign. Now, I have not always read the signs correctly. Note the rainbows in New Zealand in 1985.

But I am taking this as a good sign. Now we will see what happens.

'Russell? Colin Russell?'

'That's me, fella.'

'You come now.'

Colin – the 59-year-old Australian radio operator – stands up in a holding cell on the ground floor of St Petersburg's Kalininsky District Court, where his lawyers are about to argue for his release on bail, and the Russian state will petition for his continued detention on a charge of hooliganism. It is the morning of Monday 18 November and seven of the crew have been brought to two courthouses in white *avtozaks* with blue stripes down the bonnet and flashing blue lights on the roof.

Colin puts his hands behind his back and turns around so the guard can cuff him. Then he steps out of the cell and he's led away by three young officers in starched blue uniforms and fur hats. As they round a corner and approach the courtroom, a crowd of journalists raises cameras and microphones, and the air is filled with clicking and questions and shouts from supporters.

In the Arctic 30 campaign hubs in Moscow, Copenhagen, London and Amsterdam, and on laptops around the world, nervous colleagues, friends and family are watching the live video link broadcasting from the courthouse.

Two days ago there were demonstrations in 263 cities in forty-three countries around the world, calling for the release of the crew. In India there were thirty hours of protests across thirty cities. In Germany, huge lantern-lit marches to

Russian consulates took place in Berlin, Hamburg, Munich, Frankfurt, Bonn and Leipzig. More than thirty thousand Russians have submitted official complaints to the authorities against the jailing of the Arctic 30, putting themselves at risk of retribution.

Colin smiles. He's still wearing the battered blue sweat-shirt he had on when the ship was seized nine weeks ago. His grey hair is combed back, he's wearing a thick beard and steel-rimmed glasses that rest on the end of his nose.

'It's all good,' he says to the supporters, who are jostling with journalists in the hallway. 'All good. Hello, everybody.'

The guards direct him through a door and a moment later he's being uncuffed and locked in a cage in the courtroom. A female judge in a black robe is sitting in a huge leather high-backed throne behind a table on a raised platform, the Russian flag draped from a pole to her side. She is young, late thirties maybe, with tied-back dyed blonde hair and black thick-framed spectacles. In the corner a bank of TV cameras is facing the cage, maybe twenty-five journalists. There's a row of spider plants along the wall below a broad window. A young female guard all in black, long hair under a baseball cap, stands in the very middle of the courtroom, between the media and the prisoner.

The campaign hubs have people stationed at court using Skype to keep them updated on the multiple simultaneous hearings.

Jan Beránek: UPDATE: Colin just arrived at the courtroom.

Jan Beránek: UPDATE Colin's hearing: Investigator
argues that reasons of detention are still active,
so IC asks for detention. 'Colin could disturb the
investigation and hide evidences.' He asks for a
detention until 24 February.

Jan Beránek: UPDATE Colin's hearing: Prosecutor
agrees with investigation. Based on Leninsky court
decision, the detention is legal. Prosecutor says Colin
could escape.

The prosecutor claims there was a violent attack on the oil
platform and that Colin should be held for at least three
more months while the investigation continues. He says
Colin needs to remain in jail because he could flee Russia
and escape justice.

Jan Beránek: QUOTE from Colin's hearing:
Lawyer: 'Will you escape the court?'
Colin: 'I'm innocent, I have nothing to run from.'

The judge looks up. She folds the file closed and announces
a recess while she considers her judgement, then she leaves
the courtroom. She retires to her chambers, where normally
in the course of Russian justice – such as it is – a judge
might check their emails or perhaps call Moscow to find out
what the ruling in their hearing should be. Thirty minutes
later she's back.

Jan Beránek: IMP UPDATE Colin's hearing: The judge
is now reading the verdict. With very weak voice.

Jan Beránek: Get ready to contact Colin's relatives.

She's mumbling through the words in barely audible Russian.
Thousands of miles away, on every continent bar Antarctica,
families and campaigners are staring at their screens. It's
coming. Any moment now.

Jan Beránek: IMP UPDATE COLIN: Verdict =
DETENTION UNTIL 24 FEB

Jan Beránek: No bail. Detention.

Colin shrugs and shakes his head as hearts sink in cities across
the world. He gets to his feet and makes a V sign for the
cameras, which by now are clustered around the cage. And he
says, 'Thank you to the world for coming to our cause. I want
to thank you. Everybody around the world, thank you, thank
you. You're all beautiful people.' His face breaks into a grin. 'I
love you all. I love everybody. I am not a criminal.'

The fur hats cuff Colin's wrists and lead him out, and as
he's taken down the corridor, with supporters clapping as
he passes, he says, 'Thank you, everybody. You're all good.
All good.'

At SIZO-5 Sini is in a meeting room waiting for a guard
to take her back to her cell. There's a little window in the
door and she can see Alex in her purple ski jacket going into
a meeting room with her own lawyer. Sini waves at her and

299

Alex waves back. Then Sini sees that Alex is crying. Alex's lawyer comes to the window and shows Sini three fingers and mouths the words, 'Three months. Colin. Three months.'

Alex wipes her eyes. Sini is making a shape with her hands. Alex squints to see better. Sini has made a heart with her fingers and she's mouthing the words, 'I love you.' Sini quickly scribbles a note and gives it to the translator, who walks through to the other room and passes it to Alex. It says, *It will be okay. Be strong. I love you.*

Dima is watching the television news. He stares at the screen, at pictures of Colin. He shakes his head as the reporter explains the Greenpeace prisoners are staying in jail, then he sits down and writes a letter to his son Lev on the Pacific island of Vanuatu.

Dima has put off this moment for as long as he could, but now he knows it's at least three more months, and possibly years in jail. He knows Lev will take the letter well, he's a good kid, he has some of his grandfather inside him, and a lot of his great-grandfather and namesake, the famous dissident Lev Kopelev.

As he holds the pen above a sheet of paper, Dima's mind slips back many years to when Lev was just a kid, maybe thirteen. The family was living in Sweden and one night there was a power outage. Lev couldn't play computer games so he disappeared outside with his friends. At 2 a.m. he still hadn't returned. Dima was frantic. Then at 3 a.m. *bang bang bang* on the door. Two burly policemen stood there with skimpy little Lev between them. Anitta broke into tears, Dima asked what was going on. The cops said they found him on the roof of his school, they said Lev had talked back to them, some wiseguy stuff. Dima asked Lev, 'Did you? Did you speak back to the officers?' And Lev said, 'Well, I just said to them that as long as police violence is portrayed as justice then the justice of the proletariat is going to look like violence.'

Dima smiles at the memory. He bends over the paper and starts to write.

As you will have heard, I ran into a bit of trouble in the motherland. Been here a couple of months, and it looks like I'll be staying a bit longer. Question is how much longer. There is a risk that it will take quite a few months. And if things don't improve, years. Now, I personally am fine. It's no Club Med, but it's no Auschwitz either. My biggest concern is Mom and Luke. I'm really worried that if things extend on she'll REALLY get fucked up. Already she looks (on a newspaper foto) very skinny and depressed. Her messages to me are sounding desperate. SO, I know you got a travel plan but I wonder if I can ask you to change it? I am going to a hearing this week that should extend my imprisonment to February 24th. Who knows what happens after. It would be very good if you would consider coming home for a few months to be with your mom. Of course we can help to pay for your ticket. Sorry to dump this on you, but as you understand it's a bit of an extraordinary situation. Please don't tell mom I've asked you to do this. Anyway I hope you're having an EXCELLENT adventure so far. Hugs hugs and love to Nigh-Nush. Love you tons. Hugs and kisses from the Gulags.

Your pirate Dad

In the cell next door, Pete Willcox opens his diary and pulls the lid off his pen.

> *Three month extensions for detention. My heart did a nosedive, and I crawled up on my bunk and stared at the ceiling, which is only three inches away. Totally bummed.*

In the Room of Doom the campaigners are chewing their lips and staring into the middle distance, their eyes are red and wet. When they try to smile their lips quiver and they fight back tears. Nobody is saying much. Then a Skype message lands from St Petersburg.

> Jan Beránek: IMP UPDATE from Katya's hearing: Prosecutor said he is NOT against a bail in case of Ekaterina. The judge interrupted the hearing so that she can prepare the decision.

Katya Zaspa is the ship's 37-year-old doctor from Moscow.

> Jan Beránek: UNCONFIRMED: Katya's lawyer got information that her bail will be accepted. But we still need to wait for judge to pronounce her decision.

Mads Christensen unmutes the video link from Copenhagen. 'Did you guys just see that?

Ben Ayliffe looks up. 'Jan's Skype message?'

'He says Katya's lawyer's been told she'll get bail.' Christensen leans back in his chair and runs his hands through his hair. 'Holy shit, if that's true … '

There's no livestream to Katya's hearing so the campaigners stare at their screens, dozens of them across the globe, waiting for an update on the Skype group. A minute passes, then another. Then …

Jan Beránek: CONFIRMED: Katya's bail accepted! Now official.

The bunker erupts. In Copenhagen Mads Christensen throws his head back and thrusts his arms into the air in triumph.

Jan Beránek: As far as I understand from the lawyers, we now have 48 hours to deposit the 2 million rubles. After that, she can go free.

Jan Beránek: IMP UPDATE from Ana Paula's hearing: The prosecutor supports the bail! Decision will only be made at 11am tomorrow.

Jan Beránek: UPDATE from Denis' hearing: Denis says that if the injustice will be extended, he will respond by a hunger strike.

Jan Beránek: Denis gets a bail as well!

By early evening, Russian TV news is running images of Denis smiling, being led out of the courtroom to cheers. It's hard to make out what's happened, but most of the activists understand that Katya and Denis got bail and the prosecutor isn't opposing release for the Brazilian activist Ana Paula Maciel, but her case resumes tomorrow. At SIZO-5 Faiza is told by her lawyer that the Russians will be freed as soon as the bail money is paid. 'But if we're getting out then why did they keep Colin?' she asks. 'Are they going to make a distinction? Because it doesn't make sense that they picked out Colin.'

'Look,' her lawyer replies, 'the phone call didn't come through in time. Colin's judge didn't get the call. In Russia we call it *telefonnoye pravo*. Telephone justice. And the phone call came late. Soon enough we'll know either way. It all depends on Ana Paula tomorrow morning.'

TWENTY-EIGHT

Frank Hewetson's diary
19th November
Two months gone since FSB Alpha Team landed on Arctic Sunrise and had us imprisoned. Just saw the morning news and Anton translated that Denis + Katya have been given LIBERTY!! Great news. Saw pix of Colin and Kieron too but no info on them. It's now Arctic 28!

Pete Willcox's diary
19th November
Just saw the news. Katya, Colin and Denis, in prisoner cells in court. They all looked upset. Obviously we are all being detained again … I got some veggie food this morning. Smells like shit on a shingle. Have not heard from Dima yet. Don't trust, don't fear, don't beg.

Ana Paula Maciel, a 31-year-old biologist from Brazil, is brought to court with Sini and Camila and taken immediately to the courtroom for her hearing. Sini and Camila shuffle up next to each other in the holding cell and squeeze each other's hands.

Jan Beránek: UPDATE from Ana Paula's hearing: Ana
is in cage, but shows several slogans on paper sheets.
It says 'Save the Arctic', another one 'I love Russia but
I want go home'.

Jan Beránek: VERDICT on Ana Paula: YESSSSSSS! We
are getting a bail!

Mads Christensen punches the air. The Room of Doom is a
riot of cheering and whooping. Then suddenly Ben Ayliffe
cries, 'Whoa, wait! Look.' He's pointing at his laptop screen.

Jan Beránek: STOP STOP

Jan Beránek: HANG ON

Mads: Roger, standby on Ana Paula

Jan Beránek: MAY BE OPPOSITE

Jan Beránek: CHECKING WITH LAWYERS

The room is silent except for the ticking hands of those three
clocks above the hand-drawn flags. A minute passes. And
another. Nervous glances are shared. Then a new message
drops onto their screens.

Jan Beránek: Yes, bail confirmed for third time.

Jan Beránek: It IS bail.

And now the room erupts with applause. The video screen is filled with boxes showing arms aloft in campaign hubs around the world.

Outside the holding cell in St Petersburg a guard appears. Sini and Camila look up. The man stands back and Ana Paula walks in, head bowed. She's stood in the middle of the cell, her face a painting of grief. 'You were right,' she says, looking up. 'You were right, Sini, it didn't happen. Three more months.'

Sini and Camila reach out to touch her, but Ana Paula can't hold it in any longer, she jumps up on her toes, her face breaks into a huge grin and she screams, 'We're outta here, girls! We're gonna be *free!*'

'What?'

'It was a joke! I got bail.'

The corridors of St Petersburg's Primorskiy courthouse echo to the distant sound of elated screaming.

Jan Beránek: UPDATE from Sini's hearing: She was just brought into the courtroom.

Jan Beránek: UPDATE from Camila's hearing: Her quote '*ya tuvieron suficiente tiempo para darse cuenta que soy inocente nada mas*' / they had enough time to understand I am not guilty.

Jan Beránek: VERDICT on Camila: Bail granted!!!

Jan Beránek: UPDATE on Sini's hearing: Judge went away to prepare verdict.

Jan Beránek: Sini's final statement: 'There are no
reasons to keep me in detention. There were no
reasons in the past two months either. And I will never
stop believing in the right things.'

Sini still doesn't think they'll give her bail. She thinks they're
going to single her out because she climbed the oil platform,
so she uses the hearing to unload on the prosecutor and the
investigator. She comes to the front of the cage and jabs her
finger at the investigator, saying, 'You're a liar. You tell lies.'
Then she points at the prosecutor and says, 'And you. Yeah
you. You're hiding evidence.' She wants to show them she's
not scared, but her lawyer looks back at her and hisses, 'Sini,
maybe you should say a bit less.'

The judge retires. Half an hour later he returns with his
verdict. Sini braces herself. The judge is speaking now, the
translator is whispering the words to Sini.

' … and therefore the detainee is released while the investigation continues.'

Jan Beránek: VERDICT ON SINI – bail granted.

Sini looks up. She's staggered. Bail? She got bail? It's hard
to believe. She's actually going to be freed? Her lawyer looks
back and sticks her thumb up. A smile breaks on Sini's lips,
she wants to leap and jump and hold somebody.

Maybe, finally, it's over.

By the end of the day, twelve of the thirty have been
granted bail. They just need the two-million-rouble bonds
to be paid, then they'll be free. At Kresty, Dima is watching

Vasily make an extraordinary tomato relish and caviar sandwich when he hears the word 'Greenpeace' on the television. He looks up and sees Camila's face. And he knows.

That night across the SIZOs most of the activists watch the TV news in their cells. They can see the smiling faces of their friends. They know they're getting bail. They punch the air or scream ðut through the window or hug their cellmates, or they simply lie on their bunks and close their eyes and bathe in relief.

Dima runs his thumb under the seal of an envelope and pulls out the letter to Lev. He reads it through, then thinks momentarily of ripping it up and throwing the pieces out of the window. But then he stops himself. No, he's not yet free. They might still single him out. Him and Pete and Frank. He remembers the faces of the competent authorities – Gerbil and Helmet-hair. They were FSB goons, up from St Petersburg. He's sure of it. He's in their home town now. His bail hearing is on Friday. This is only Tuesday evening. Anything can happen in the next three days. He slides the letter back into the envelope and puts it under his pillow.

Frank Hewetson's diary

20th November

Well. Fairly major surprise meeting with Pavel [his lawyer]. Guard came and got me and took me to the meeting room where they were waiting with big smiles. 12 non-Russians have effectively been given bail, including SINI which has major implications for those of us in the rhibs.

[later ...] 6pm news showed Anne, 3rd mate, getting bail. Really encouraging news. My questioning of Anton about the news report was a bit enthusiastic I think. He finds it hard not to show his own feelings of jealousy. Somehow I must show restraint on each release of an Arctic 30.

The next day nine more are brought to court, including Alex. There are so many photographers that it feels like walking on the red carpet, except she doesn't feel glamorous because she's just stepped out of prison. There are flashes going off, people barking questions. 'Do you have a few words to say? Alex, how do you feel? Do you think you'll get bail? Alex? Alex?' Then she's pushed into the cage. The judge asks her if she has anything to say. Alex tells the court: 'Every day in prison for me is connected with the struggle. I joined Greenpeace because I care about the environment and think the Earth should be saved for future generations.'

The judge retires then returns with the verdict. The translator is whispering the words in Alex's ear and she's thinking, hurry up and just spit it out, please. And then the judge gets there. She says Alex has got bail.

She feels light-headed. Delirious. It's like she's become detached from her body and is watching herself in the cage, getting to her feet and smiling for the cameras. The whole courtroom is filled with reporters. 'Alex, Alex. What's your reaction? How do you feel? Have you got a message for your parents? Alex?'

Frank Hewetson's diary

21st November

The steel section masquerading as a bed base is starting to hurt my hips. And ribs. The mattress has inexorably sunk to the profile of the steel plate and is ridged. I think I also pulled a muscle in my back at gulyat playing hacky-sack with Anton.

[later ...] Back from the court with bail conditions granted! Pavel told me I won't get out of Kresty till Monday as payments are taking time to process. I'd love to be out by Friday but no go. Media presence was huge for such a small room. A mad scramble with police getting very angry with SKY NEWS team. They fully got into a scrum. Told me I was live and I did a piece. I also gave my speech about the FSB being the armed hooligans that morning, not Greenpeace. Stewey [head of the media team] was in the media pack! Great to have seen him. Really feels like a major step on the road to freedom. In the holding cell at court I saw a Column 88 stamp with swastikas each side + Russian lettering underneath. Lots of right wing thugs here I guess. I wrote 'Nazi Scum' above with an arrow pointing down.

[later ...] Just got taken to the phone room and allowed to call home. Feckin cards only last 3 minutes and the system failed 80% of times but I got to talk to Nina and then Nell. Everybody cried a bit. She's really really missed me and must have been hit quite hard by the whole thing. So so good to have got just those few brilliant moments though. My darling daughter. Nina was so lovely to talk to. She has been the best wife possible during these 2 months.

The campaigners are hoping the first of the activists will be freed soon. They expect it to be one of the Russians – Denis or Andrey. But at SIZO-5 a lawyer is clutching a receipt slip for the payment of a two-million-rouble bond, and she's refusing to leave without her client. 'This is nonsense,' Ana Paula's lawyer tells the governor. 'Just release her. This procedure is crap. The judge gave her bail, we've paid the money. I won't leave without her.'

And suddenly, unexpectedly, Ana Paula Maciel from Porto Alegre in Brazil is told by a guard that she's being freed. Half an hour later, from a grey door down the side of SIZO-5, she emerges into a Russian winter's evening, surrounded by journalists, blinking into flashbulbs, grinning with a combination of delight and shock, holding in her hands a sheet of paper on which she's written the words 'SAVE THE ARCTIC'.

Pete Willcox's diary

21st November

I got bail. I am still in prison but expect to be out tomorrow or Saturday or Monday at the latest. Ana Paula's hearing was Monday and she was out yesterday afternoon.

I was up and down all day. Finally about 3.30 (?) I came back to hear the judge (young blonde woman – mid 30s?). The judge started talking and after a bit the translator tries to keep up. Well the judge lists all the reasons why I should not get bail. It goes on for 5 minutes, but by the middle I am completely trashed. I tried to figure out why I am not being bailed, and the only reason I can think of is that as captain, they want to make an example out of me … I start to get myself mentally prepared for more prison. Then the judge starts saying all the reasons why I should be bailed. And it finally dawned on me that she had just outlined the prosecutor's case. So my hope sprang up again. After a few minutes, my lawyers both started giving thumbs up. And then the judge says it real plain ... bail is granted. I sighed, closed my eyes and looked down to the right, and an absolute blast of motor-drives went off. I was too drained for even a smile.

But now Pete is worried. He's back in the cell and his mind is racing. He's convinced that the others are blaming him for all this. He's been questioning his judgement as a captain since commandos seized his ship. Shouldn't he have got out of there as soon as the coastguard started firing warning shots? Some of the crew were kids, they never signed up to years in jail. Pete imagines them lying in their cells, waiting for release and working out how they ended up in this mess in the first place, and deciding it was his fault. They'll be out soon and they're going to have their say. They're going to want to unload, and it's all coming his way. And maybe they're right, he thinks. Maybe they're right.

At 2 p.m. Camila is told her time has come. She's taken to the meeting room. She waits alone for five minutes, then the door opens and Sini appears with 26-year-old Danish sailor Anne Mie Jensen. They're taken to the main door of the prison. They're standing before it now. Then it swings open, and another door opens in front of it, and beyond that is a bank of cameras below a broad sky. Camila walks out. Then Anne Mie. Sini takes a step forward but she doesn't realise the last door is so heavy. It swings back, Sini leans into it but she can't hold it back and she's momentarily squeezed between the prison doors in front of the cameras. She pushes again, steps into the open air, holds her arms aloft and shouts, 'I'm free!'

It feels like being born again. She wants to run somewhere, anywhere. And she wants to hug anything with a heartbeat. A minute later they're in the lawyer's car, speeding down a St Petersburg street.

'Is it okay if I open the window?' asks Sini.

'Yes, sure. Open it.'

So she opens the window and sticks her head through it and shouts, 'I'm free! I'm free!' Drivers in passing cars turn their heads, people walking on the pavement look up, they see blonde hair flapping and a woman screaming into the wind.

'I'm free!'

The women are taken to the Peterville hotel in central St Petersburg and given keys to their rooms. In a third floor corridor Camila is fiddling with a key in the lock while Sini is next to her, jumping up and down on her heels. Behind them Anne Mie is hugging Ana Paula.

Camila turns the key in the hotel door and cries, 'I can open the door! I can open the door! And I can sleep with the door unlocked!' She pushes through the door into the hotel room, Sini bounces after her and the two of them punch the air and whoop. They run to the window and disappear behind the curtains. The fabric jerks and bulges as they wrestle with the window fixture, then the window opens and Sini shouts out into the street.

'Freedom!'

And Camila cries, 'We are free!'

Then Camila's head pops out from behind the curtains and she says, 'Hey, maybe we should do a banner hang here.'

And a voice from the corridor says, 'You can't do a banner hang, you only just got out of jail!'

Sini jumps onto one of the beds and Camila joins her. They're bouncing from bed to bed, whooping and screaming until Sini is laughing so much she falls off and rolls onto the floor and Camila collapses onto her back and shouts, 'I feel like a three-year-old child!'

More people are piling into the room. A bottle of champagne is produced. Sini says she hasn't drunk alcohol in five years. 'But this is the day to break it.' She takes the bottle, pops the cork and everybody's arms go into the air in celebration, and just at that moment the Italian activist Cristian D'Alessandro waltzes into the room sporting a grin the width of a dinner plate.

'I just took a shower,' he says, 'and I'm a free man!'

TWENTY-NINE

Frank Hewetson's diary
22nd November (Friday)
THAT'S IT – JUST GOT GIVEN 5 MIN WARNING
– I'M BEING RELEASED …
   Turma racing over.

Frank feels joy surging through his body, to the tips of his fingers and toes. It washes over him like a cleansing shower. He takes a deep breath, smiles and starts packing his bag, but as he glances at Anton his cellmate looks back at him and Frank can see he's wishing it was him. Outside his cell the governor of Kresty is standing in the corridor, watching him through a gap in the open door. The man says something in Russian to the guard, and the guard turns to Frank and says, 'You put in a request to see the cathedral?'

Frank looks up.

'What?'

'You asked to see the cathedral.'

'Yeah. Yeah I did.'

'You are religious?'

'I'm a Catholic, sort of. But I was so amazed by the architecture here. You have to understand that my cell window in

317

Murmansk looked out over razor wire and a guard tower. Here it looks out onto this wonderful dome.' He grins. 'Shame I'll never get to look around.'

'But you put a request in?'

'Uh huh.'

'The governor, he wants to know if you still want to see it?'

Frank stands up straight.

'The cathedral?'

'Yes.'

'Well I'd love to, my friend. But I'm afraid I'm leaving you now.'

'You want to go, before you leave?'

Frank glances at the governor. He wants to get out of this place as soon as he can, it's finally happening, some nights he thought he'd be in prison for fifteen years and now he's standing on the cusp of freedom. But still he hears himself saying, 'Okay, sure, yeah, let's go see the cathedral.'

'Finish packing. I take you.'

Frank stuffs his possessions into the bag, zips it closed, swings it over his shoulder and nods. 'Let's do it. Let's see the church.'

They walk him down hallways and up stairways until they're standing before a grand wooden door. He walks through it, then they pull the door shut behind him and let him wander around, all alone. He's surrounded by icons. A gold altar. Huge silver candelabras. And all inside a prison, like booty. It's absolutely beautiful. Stunning, he thinks. He can see the faded colours of a fresco. He looks up and sees the entire working of the bell system. The silence is odd. The size of the space feels strange to Frank. He hasn't felt space like this for months. It's disorientating, like it's hard to balance and he could fall over. It smells of dust, decay and dry, rotting timber. Every so often the door creaks as the guards look in and check he isn't stealing something. And eventually they tell him his time is up.

The moment has come for Frank Hewetson to leave jail. He draws a deep breath and nods. He takes a last look, runs a hand through his blond hair then strolls over to the guards. They pull the wooden door closed behind him, he slaps one of them on the back and says, 'Okay, brother. Let's do it.'

Meanwhile, Pete Willcox is in a car on the way to the hotel. And he's worried. He thinks he's about to be blamed for what happened. He's as nervous as he is excited. Nervous about meeting the others. A captain's first job is the safety and welfare of his or her crew. By sailing them to the *Prirazlomnaya* oil platform he thinks he nearly screwed up the lives of thirty people. The car pulls up outside the Peterville. He throws his bag onto his back and climbs out. Sini, Alex and Camila rush forward and throw their arms around him.

'Pete, we missed you.'

'We love you, Pete.'

It's one of the best moments of Pete Willcox's long, eventful life.

Back at the courtroom the final bail hearings are wrapping up. The last of the activists to be brought to court are Dima, Roman and Phil. In Irvington, New York, Pavel Litvinov watches footage of his son standing in a cage addressing the judge in St Petersburg. Forty-five years earlier, Pavel stood up in a similar cage in a Moscow courtroom and delivered a speech that resonated across the Soviet Union.

'There was never any question for me whether I would go to Red Square or not,' he told the judge that day. 'As a Soviet citizen I deemed it necessary to voice my disagreement with the action of my government, which filled me with indignation ... This is what I have fought against and what I shall continue to fight against for the rest of my life.'

Now his son is standing in a courtroom cage before another Russian judge. Despite bail being granted in the earlier cases, the Investigative Committee is forcefully arguing that Dima should remain in jail in case he absconds. Could Helmet-hair and Gerbil have arranged for him to get the same treatment as Colin? Dima grips his hands around the metal bars and clears his throat.

'We sailed into Arctic waters in September to bear witness,' he tells the court. 'To tell a story. We came to tell a story about the risk to the global climate and the Russian environment from opening up Arctic oil fields to companies such as Gazprom, Shell, Rosneft, Exxon and others. It would be unthinkable and irresponsible of me not to use the opportunity of this criminal trial to tell that story now. I want a trial. I want to be able to speak to the authorities and society of Russia, to the global media, to explain that our journey to the *Prirazlomnaya* was not an act of hooliganism or disrespect for society. Instead we were sounding an alarm, a warning of imminent danger.'

Camera shutters click, TV crews jostle, Dima squeezes the bars of the cage and stares squarely, defiantly at the judge.

'I was proud of him,' says Pavel. 'I was proud of everything he did. I felt we gave him the right values and I loved that he was devoted to something more important in his life than making money.'

The judge retires for half an hour then returns to deliver the verdict. Dima is granted bail. He lowers his head in the cage and nods. Roman and Phil are also told they will be freed. They're all taken back to Kresty, where Dima and Roman are told to prepare for immediate release. But Phil's paperwork hasn't been completed and he won't be freed yet. Dima packs his huge pink bag, throws it onto his back and picks up his towel. Stitched into it are the words 'SAVE THE ARCTIC'.

The guards take him down the hallway, where he meets up with Roman. Then they're taken to the governor.

'I have a gift for you,' the man says, pushing something into Dima's hand. Dima looks down.

'A fridge magnet. Wow.'

'Made in Kresty,' says the governor. 'It has a picture of the prison on it. And the cathedral.'

Dima nods uncertainly. 'Nice. Thanks.'

'And look,' says the governor. 'It has writing on it. A message.'

Dima brings the fridge magnet up to his face. He pushes his circular steel-rimmed spectacles up his nose and focuses. Stamped along the bottom he sees the words 'Freedom to the freed and heaven to the saved'.

'That's … that's fantastic. Thank you.'

Dima and Roman are taken to a counter. They're just minutes away now, they just have to sign for the possessions that were taken from them when this whole thing started. Then they're taken over to a steel door. On the other side is freedom.

It's a sea of cameras, bulbs flashing, shouted questions. Dima holds up the towel and tells the throng of reporters, 'It is wonderful we are out but it is still not over and we have a lot to do. They say the *Prirazlomnaya* will start drilling in December so we haven't won yet. It's just one step towards victory. But it feels so good to breathe the air of freedom, no question. We are out so it is a step forward for justice. But we know that the Arctic is still under threat, so the struggle will continue until we stop the destruction.'

In Hamburg, the campaign's legal chief Jasper Teulings is sitting in the public gallery at the International Tribunal for the Law of the Sea, waiting for the ruling. He's nervous, the room is packed. He's sitting next to Kumi Naidoo, prodding him in the ribs as the Japanese chair reads out the ruling in English.

' … the Russian Federation shall immediately release the vessel *Arctic Sunrise* and all persons who have been detained upon the posting of a bond or other financial security by the Netherlands.'

Did Putin know the ruling was coming? Is that why they're already free? The choreography of the moment seems advantageous to the Kremlin – Russia's oh-so-independent judicial system freeing the thirty before a tribunal of foreign judges forces the issue.

They're nearly all out now. They're pouring into Fabien Rondal's room to lay their hands on phones, laptops, anything that allows them to contact their family and friends. But Faiza Oulahsen – the 26-year-old climate change campaigner from Amsterdam – is stood on the pavement outside, alone. After so long in isolation she's finding the wave of unfamiliar experiences uncomfortable. Lights, traffic, three people talking to her at the same time, the fact that she can actually make a phone call. When she speaks to her family and friends she tells them they're not allowed to call her, she'll call them when she can cope with it. That first day she feels like isolating herself again. She needs space and peace to adjust.

It is said that one's alcohol tolerance diminishes with abstinence, but the tolerance demonstrated that night for Russian beer and vodka and Georgian wine is quite extraordinary. They inhale booze all night as they cluster in groups, holding hands and sharing stories.

'Did you meet Popov?'

'Oh, he was such an arsehole.'

'He was crazy. He hated me. He had this thing about my potatoes.'

'Your *potatoes*?'

'Seriously.'

The bond that has grown between the women – especially Sini, Camila, Alex and Faiza – is striking. They hardly knew each other before they were jailed. They'd only been together ten days on the ship when it was stormed. Now they have their arms draped around each other's shoulders, and they look inseparable.

As the party rages around him, Dima calls a number in Lexington, New York.

'For so many years we were both very reserved people in different ways,' says Pavel Litvinov. 'At least towards each other. So it was one of the best conversations I had with my son for many years. There was so much warmth. We were always good friends, we always knew what was happening with each other, but we might not talk for many months. The last time I'd seen him was in California with my daughter, his sister, and we were happy to see each other, but then I would go to New York and he would go to Sweden and we hadn't talked for many months. So it was so wonderful to talk to him, and such a relief. I told my wife that it was all totally worth it because I had the warmest conversation with my son for many years.'

THIRTY

Twenty-eight down, two to go. Only Phil and Colin are still in jail. Phil's paperwork wasn't completed in time on Friday afternoon and he was told he'd have to wait until Monday.

The freed activists spend their time sharing stories or speaking to husbands, wives, boyfriends, girlfriends, kids, parents and journalists. So many journalists. But they need to be careful. Their legal ordeal isn't over yet. The lawyers say they're still charged with a crime that carries seven years inside, the investigation hasn't been dropped, they're already being given dates to report to the regional Investigative Committee headquarters where they'll be interviewed by senior officers.

This is only over when they get home.

When they're not in their rooms they mingle in the café downstairs. Most aren't ready to venture far from the hotel. Even when the café is largely empty, the other tables are often occupied by one of several middle-aged Russian men who all wear leather jackets and have Bill Gates haircuts. They love their iPads, these men. Always fiddling with them, holding them up, pressing the screen, taking photographs.

Pete notices scratch marks in the wood around the lock on his door. He's pretty sure they're new. In Frank's room the plug sockets stop working, he goes downstairs to complain, reception phones a number and a moment later a man in a shiny leather jacket slips out of a cubicle in the lobby and skips up the stairs. Frank darts away and follows him. He watches the man walk into his room. Then, perhaps sensing he's been spotted, the man exits Frank's room and proceeds to walk into an adjacent broom cupboard and close the door. Frank waits. He can hear rustling from the other side of the door, the sound of switches being flipped, then the door opens and the man walks out. He nods at Frank, brushes past his shoulder and disappears around the corner.

Three days after Phil was told he wasn't yet free, the activists are stood on the pavement outside the Peterville hotel. A car pulls up. Phil climbs out and holds out his arms, raises his eyes to the sky and says, 'Look. Look at this. It's a sky. A sky. And there's nothing in front of it. And … can I just do something that I haven't done? I'm coming back but … ' then he tears away and sprints down the pavement, and keeps running, and keeps on running until he's lost in a crowd of Russian shoppers.

*

Oh what a load of shit. I hope I get bail. It would be unreal if I had to stay in this place. It would not be fair at all and that is for shit shoot and sure, but then again who am I? I am a nobody here. Nobody.

A week after his friends were released, Australian radio operator Colin Russell is appealing the extension of his detention. He's taken to a cell and sat in front of a video-link screen, on which he can see a courtroom and his own face. As the hearing proceeds, he jots down his thoughts in a notebook, minute by minute.

I'm guilty till proven innocent and that is the short end of the stick, I think. But I cannot predict what will happen. The fuckers have asked that I stay longer. And there's heaps of cameras and I'm not sure if they will help me or not.

I look at the court and I see three chairs, and there's only one judge. There's only one judge though, so it must not be too important. The press gallery is huge. Once again please, please, please, please grant me bail.

Universe, this is not fair and I'm in this joint. I need to be with my family, I need to be with my friends. We will see what happens I guess.

The judge addresses the court in Russian. Colin can see the moment of truth is coming. The judge stands up and leaves the courtroom, the audio on the link cuts out. All he can see is an empty high-backed black leather chair behind a bench.

The judge has been away for some time now so he's thinking hard or shagging the clerk of the court on the desk in there. I wouldn't put it past him. At least the courthouse looks clean and not like the last one.

The judge returns. He settles into the chair, lifts a sheet of paper and begins reading. He stops speaking and lays the paper on the desk. In Colin's ear the translation catches up.

I'm free.

Despite the ruling of the ITLOS court, the Kremlin is still refusing to allow the crew to go home. The four Russians are allowed to be with their families in Moscow, but the others must stay in St Petersburg. Increasingly it feels like being stuck in an airport departure lounge where the planes never leave.

The biggest complaint from the crew is that they have to share rooms at the Peterville. After months in jail they want to choose when to have company and when to be alone, so everyone moves across town to the Park Inn Pribaltiyskaya, an ugly brown Brezhnev-era monolith that was built for a Soviet conference in the seventies.

Three weeks after their release they're still confined to the city. Every few days they're summoned to the Investigative Committee, where senior officers grill them with the same questions. *Who was in the boat? Who was in charge? What part did you play in the terrorist attack on the oil platform?*

Some of the thirty are becoming settled here, enjoying their freedom and each other's company, piling into a hotel room in the evening to party until morning. But others are on the periphery of the group, factions are forming, the pressure of confinement and the uncertainty around the legal case means low-level bickering sometimes breaks into outright arguments.

And that question of blame is still raw. A couple of guys still believe Frank and Dima didn't do enough to warn them how heavy things could get. Sometimes in the bar there's a table of former prisoners chatting loudly and downing beers, with Frank sat alone in a corner of the room, reading a book.

The pressure is building. In his diary Frank describes the atmosphere as 'draining'. He says the 'continued predicament of uncertainty and group squabbles have led to a feeling of despondency'. He talks of 'meals, meetings and mini-meltdowns'.

In the long legal summits between the lawyers and the thirty, the tension is thick. In one meeting Dima says, 'I hope we're still in Russia for the Olympics. If that happens we'll make a huge stink out of it.' And Pete throws his head back and cries, 'Jesus, Dima. Are you fucking *insane?*'

Frank Hewetson's diary

10th December

So so very good to finally grab both Nell + Joe in my arms on Friday night at the bottom of the stairwell. I was in floods of tears. My lovely boy + girl. I missed them so so much. So good to sit down all 4 of us and argue, giggle, snort with laughter and see Nina go the whole way from giggling to crying as she does at emotional times. So lovely to be back with my loved ones. So good.

Hopes are turning towards an Amnesty Bill passing through the Duma – the lower house of Russia's parliament. The amnesty is scheduled to coincide with the twentieth anniversary of the adoption of Russia's constitution in December 1993. The Bill is set to free many thousands of prisoners across Russia, but at its first reading in the Duma it becomes clear that in its current form it won't apply to the Arctic 30. The text covers hooliganism – the crime the crew are charged with – but only for people already convicted. Pussy Riot will qualify, but not Greenpeace. Powerful figures in the Duma indicate their support for a broader amnesty, but ultimately it's for Putin to decide if the saga will now end.[116]

One week before Christmas Day the Amnesty Bill comes back before the Russian parliament for a final vote. At the last minute an amendment is inserted to include not only those convicted of hooliganism, but also those accused of it.[117] As things stand, the Arctic 30 are now included. At 4 p.m. Moscow time there's a vote in the Duma. About half the activists are clustered around tables in the hotel bar, watching the debate on computer screens. When the result comes through, it's resounding. The Amnesty Bill is passed. But the moment feels oddly anticlimactic.

Silence. Somebody sniffs. A few of them take sips of beer but nobody proposes a toast. This is not how anybody imagined it would be during the darkest days of their ordeal. A journalist approaches them. He raises a video camera, but there's nothing to film. He asks if they can contrive a celebration, and Sini – always keen to accommodate people – fakes a hug with Phil and breaks into an unconvincing smile. Then she sits down and shakes her head.

'They can put that fucking amnesty up their arse,' she mutters. 'I don't want to have an amnesty from Mr Putin. He was in the wrong all along. I didn't do anything I need to be amnestied for. He wants us to say thank you for his forgiveness. No way.'

The following day Putin holds his annual press conference in Moscow. It's a national event. Two thousand journalists are invited; the questions have already been decided. He announces that Mikhail Khodorkovsky, the most high-profile political prisoner in Russia, will also be freed.[118] Then he confirms the Arctic 30 will be amnestied. But he says their ordeal should serve as a lesson. Putin says he suspects the Greenpeace protest at the oil platform was an 'attempt at blackmail and extortion, or they were carrying out some-body's order to stop our work'.[119]

'Screw you,' Dima shouts at the screen. 'You don't owe me an amnesty, you owe me an apology.'

Khodorkovsky, Pussy Riot and the Arctic 30. The Kremlin is taking out the trash before the Winter Olympics in Sochi in six weeks. None of the activists is happy that their freedom comes wrapped in Putin's magnanimity, but they know they won't get home unless they take the amnesty. It's the only way to avoid a court case on trumped-up charges, then years in a Russian prison.

But Denis Sinyakov can't do it. He knows some of the leading figures in the Russian opposition movement. He was photographing Pussy Riot's protests before they were famous. He says he won't sign the amnesty. He wants to fight the case in court.

'I don't care about Putin,' he says. 'This is about my truth. My lawyer says we should go to trial and fight like lions. He says if we accept it we'll just help the investigators and the authorities to win.'

Denis's decision is an earthquake. Mads Christensen and his team have been told it's all or nothing – either thirty people take the amnesty or the politics of the moment get complicated. The campaigners think the others might not be allowed to leave Russia if anyone refuses Putin's offer. A rumour goes around that ten people have already been granted the amnesty but that the process has now been halted. People are saying the investigators aren't signing off amnesties for the others because of Denis. But when Denis and his lawyer investigate, they find that's not true.

Video conference meetings involving people on multiple continents are dominated by the Denis dilemma. People are throwing their hands up or banging their heads on the table, saying, 'Can't someone just *speak* to him?' or 'Doesn't he understand what's at *stake* here?' It's like a powerful wind is blowing the campaigners towards an outcome that suits both Greenpeace and the Kremlin, one where they both sweep their pieces off the chessboard and declare a draw. Denis gets fifteen, maybe twenty phone calls. He's subjected to pressure from every corner. He's not sure who to believe. Nobody is. But eventually, for the good of everybody, he gives in and signs the amnesty. But he feels cheated. By the investigators and by Greenpeace.

'It means we get our freedom,' Phil tells the others in the bar that night. 'But I agree with Denis. He wanted to fight it through the courts. But the thing is, you can't trust the system here. I want justice to be done and this amnesty isn't justice. They haven't dropped the fucking charges. The amnesty sucks. It feels shit, it feels like a letdown. But we have to admit, it's the news we wanted. It's the only hope we have to get out. It's the only way we're going to get home.'

'In life, everything comes and goes,' says Camila. 'Let's look at it differently. Maybe one day we'll give them amnesty, but for a crime they *did* commit.'

It's Christmas, ninety-nine days since the protest, five weeks since the Arctic 30 were released from prison. All they need now is a stamp in their passports that says they can go home.

Frank Hewetson's diary

25th December
Despite being kept waiting for 30 minutes in a cab while
[French activist] Franky Pisanu had coffee + ciggies, we
made it to the Investigative Committee in time. Process
was quick and 3 of us with our respective lawyers departed
at 12:00 with our copies of signed amnesty. This officially
halts the prosecution case against us. What a very strange
Christmas morning. Off to the migration service at 3pm
to hopefully complete the final process of this farce + get
an exit visa.

26th December
Quite a momentous day with everyone of A30 focused on
travel plans + returning home. Dima L actually leaving
tonight on train to Helsinki and then ferry to Stockholm.
I've still yet to see my travel tickets to and back from
holiday with the family. Looking forward to hugging the
whole family, running into the sea and knocking back too
many cold beers.

Dima is sat in the hotel lobby with the head of the media
team, writing a statement for the press. His train is leaving
in two hours. When he's finished writing it he pushes his
spectacles up his nose and spins the laptop around. 'Release
that,' he says, then he throws his pink bag over his shoulder,
takes his wife's hand and walks through the doors of the hotel
and out onto a St Petersburg street. Two hours later, just after
midnight, when his visa becomes valid, he approaches the
border with Finland.

The female border guard surveys his passport. She reads a letter from the Investigative Committee explaining that they've decided not to prosecute Dima for entering Russia illegally. The guard folds the letter and slides it back into the passport.

'You people are trouble,' she says.

'I'm sorry?'

'Why don't you protest in America? Only in Russia. You are against Russia.'

'Actually I've protested many times in America.'

'Huh. Well maybe you should stay there.' She holds out his passport, he takes it, she moves on to the next person. Two minutes later the train crosses the border. He's in Finland.

From that first day in Murmansk at the Investigative Committee headquarters, Dima has felt a fist of fear in his stomach. It tightened then loosened then tightened again, depending on how scared he was, but it was always there. But when the train crosses the frontier, it lets go. And for the first time he feels truly free.

As he's heading towards Helsinki, the rest of the crew are gathered in the hotel bar, coming to terms with a wide selection of potent Russian alcohol. That night everybody stays up until 4 a.m. They want to spend every last minute together. It could be a long time before they see each other again. An hour after they call it a night, their alarms go off. Alex, Sini and Camila sit on the bed in Alex's room and hold each other. It's time to leave.

At St Petersburg airport activists are dispersed across the check-in counters, about to scatter to destinations all over the surface of the Earth. Alex, Kieron, Phil, Anthony and the

British engineer Iain Rogers are flying to Paris then taking the train into London. Po-Paul is with them, he's taking a connection to Montreal. But Frank isn't. He's leaving alone, heading for a rendezvous with a family holiday.

The plane to Paris is full. Almost all the other passengers are well-dressed Russians. The aircraft taxies onto the runway then waits. Five minutes pass, another five, nothing. Nobody says much. Anthony makes a phone call, the others just stare ahead, waiting. Finally the engines fire and the plane launches down the runway. As it lifts off, Phil punches the air.

Frank Hewetson's diary

27th December
That was quite a night of celebration. Finding it pretty hard-going this morning. Had a phone call from Anthony on the plane. Said it was a media scrum at the terminal. He talked me through the TAXI and TAKE-OFF! 'See you on the other side ... ' He's very relieved to be out of there. Feels a bit weird being the last Brit left in Russia of the A30.

In Paris they board the Eurostar, destination London. It races through northern France, towns and villages fly past the window. There's a sucking sound as the train enters the Channel Tunnel, and twenty minutes later they burst out into the English countryside. The train slows as they approach St Pancras station in London. When it pulls into the platform the station manager jumps into their carriage and tells them to wait. She says there are many journalists waiting on the concourse, more than eighty, maybe a hundred.

The manager asks them to follow her. She leads them through passport control and customs. As they approach the open doors to the concourse Alex can see people lining the staircase, the buzz from the media throng is palpable. A press officer skips ahead and peers around the corner. There's a bank of cameras there, more than anybody imagined. The press officer walks back to the five activists and asks them to pause for a second.

'There are loads of journalists there. Shitloads. Are you ready?'

But before any of them can answer, Anthony has pushed past and the others are following him. They walk through the doors, the flashlights explode and their loved ones surge from the crowd with open arms.

At the station in St Petersburg, Sini and Kruso are climbing onto a train. Sini's journey home will be short, just three hours, while Kruso is going overland back to Switzerland. The media interest in Sini is huge, a scrum of journalists is expected in Helsinki, some are planning to board the train as soon as it's on Finnish soil. Kruso's never been comfortable with the media so he's booked a seat a few carriages down from her.

As the train pulls out of St Petersburg, Sini feels an odd sense of deflation. She stares out of the window, confused, searching her body for the euphoria she was expecting. Even when she presents her passport to the Russian border guards and they stamp it without a second look, she feels nothing. They cross the frontier. She's in Finland now. Russia is behind her and Sini thinks, okay, this should feel like something, but it doesn't.

I still don't feel anything.

She needs to be with Kruso. She needs to be with him and she doesn't care about anything else. Her thoughts right now are impossible to make sense of alone, without the others she was jailed with. She can't feel it without Kruso.

Sini gets up and walks down the train, pushing through doors and passing through successive carriages, until eventually she finds him. He's sitting alone, staring at the passing countryside. She falls into the seat next to him and throws her arms around him and holds him as tightly as she can, just like she did when they were surrounded by soldiers on the deck of that Russian coastguard ship, one hundred and one days ago.

EPILOGUE

Sini Saarela wants to jump. She's not sure she can do it, but that's what she wants to do.

It could be the most foolish thing she's ever done. The water's freezing cold, she knows it from the spray thrown up by the police speedboats. Even if she does it, even if she jumps, there's still no way she'll make it to the jetty. The sea is teeming with Dutch police and coastguard boats and there's a line of cops on the wharf, maybe twenty officers fingering their handcuffs and looking down at her. She'll have to swim fifty metres, maybe eighty, then haul herself up a ladder. That's what she'll have to do if she wants to lock herself to the oil pumps and stop the supertanker docking. But first she has to jump in with a rucksack full of rope on her back. She's wearing a helmet, a thick drysuit, a climbing suit, a climbing harness and a life jacket. She'd be throwing herself in without even knowing if she floats.

She looks over the edge of the RHIB at the churning water and screws her hands into fists.

It's May 2014, eighteen weeks since she left St Petersburg, and Vladimir Putin has just fired the starting gun on the Arctic oil rush. It was on a video link between the Kremlin and the *Prirazlomnaya*. He congratulated Gazprom on 'a big event' that 'marks the start of our country's ambitious plans for developing production of Arctic mineral and oil resources'.[120] Then Gazprom said the first oil from icy Arctic waters was being loaded onto a tanker that would soon set sail for the Dutch port of Rotterdam, to be refined and pumped into Europe's cars.

The *Mikhail Ulyanov* is 250 metres long,[121] the size of a skyscraper lying on its side, and now it's casting a shadow over Sini. It's only going at four knots and it barely makes a noise but it's drifting towards the jetty in Rotterdam harbour. Sini is standing at the bow of a Greenpeace RHIB. Behind her, looking over her shoulder, are Phil Ball and four other climbers – two Germans, a Dutchwoman and a guy from Finland.

A minute ago they had a plan. The RHIB was going to drop them at the bottom of a ladder at the jetty and they were going to climb it and lock themselves to the oil pumps. But then a police boat barrelled in and a cop jumped into their RHIB and pulled the kill cord on the engine. He stuffed the cord into his pocket and jumped back into his own boat, so now Sini and the others are drifting on the water between the tanker and the wharf, and their plan is a busted flush.

'But that fucking dirty oil was coming in from Russia,' Sini recalls. 'I felt very strongly that I wanted to get between the ship and the jetty. I wanted to say, "You're not bringing that shit in here."'

Sini wants to jump. Without that kill cord the RHIB can't get her to the ladder, so the only way she'll reach the jetty is if she swims it. But if she jumps she'll be alone in the water. There's no way the others are going to follow her in, they're carrying the same weight of kit as she is, and by the look on their faces they're not thinking what she's thinking. And if she jumps she'll have to swim through a constellation of police boats and get to the ladder before the cops on the jetty can reach it, and that's just not going to happen.

The *Mikhail Ulyanov* is close now, maybe a hundred metres away. Pete Willcox managed to slow the tanker's progress for a while, he was at the wheel of the Greenpeace flagship *Rainbow Warrior III* when she cut across the tanker and blocked its entrance into the harbour. But Dutch commandos broke into the bridge of the *Warrior* and seized control from Pete – 'Okay, so what else is new?' – and now the Russian ship is minutes from landing that Arctic oil.

Since she got home, life hasn't been easy for Sini. Dozens of journalists were waiting for her when she pulled into the train station in Helsinki. When she was in Russia she had no idea how famous she'd become. Now she's recognised in the street, the focus of a baffling, intense public interest.

'It got so personal about me. In the beginning I couldn't even go to the market because people were almost dropping their shopping baskets and staring at me. It was horrible. I didn't like it. There were lots of people who came up and talked to me, which is not like Finnish people. Of course it was very nice if it was positive because you could say thank you for the support. But then there was also negative feedback.'

Negative feedback. She means abuse in the street from people she's never met.

She wants to jump but she doesn't think the others will follow her in. And if she goes alone she can't hope to pull this thing off. But if she knew for sure the others would jump as well, she'd do it now.

Faiza Oulahsen is standing on the deck of the campaign ship *Argus,* 500 metres away, watching the police clear a path for the tanker's approach towards the jetty. 'And I could almost smell the thing I went to prison for, it was so close. The first tanker of Arctic oil, the very thing we were trying to stop six months earlier, and it's coming into my home country. It was misty and foggy and dark and I had a flashback to the platform and the Russian coastguard standing between the *Arctic Sunrise* and the *Prirazlomnaya*, with the authorities protecting the interests of Arctic oil.'

Frank isn't in Rotterdam. He's concentrating on his family right now, trying to make sense of what happened back there in Russia. 'I was in prison for all the right reasons,' he says. 'That's how I feel. We were crucified and taken to the cleaners, properly imprisoned, so you knew you were having an effect somewhere, and that kept me going. But you have to compare that to the nights where you convinced yourself you're looking at ten years for piracy. Then you're very vulnerable. Very worried. Very down.'

In his final diary entry, on 10 January, a few days after returning to London from that family holiday, he wrote:

Walked Nell to school today. Felt very good to fulfil that walk that I'd thought so much about in Murmansk. Walking Joe was on Tuesday and I took the long path back from Finchley Road to West Hampstead alongside the train tracks. For some reason I'd wanted to do that too. I was listening to the explorer Ray Mears on BBC Radio 4 Desert Island Discs. He mentions the driving force and reason why some survive for many days in the face of disaster. It's loved ones. Prison is not the perilous jungles of the outback but the isolation drove me to yearn and hope to be back with Nell, Joe + Nina.

Denis is back behind a camera, capturing new chapters in his country's long and storied history. After the amnesty he visited his newly freed friends from Pussy Riot, then he headed to Kiev to cover a revolution that would soon spark conflict between Russia and Ukraine. Everywhere he goes he's recognised and saluted, but he finds the attention embarrassing. 'You know Solzhenitsyn, he spent ten or fifteen years in jail, so it's a shame when people say, "Respect to you, you're a prisoner." It's bullshit. It was only two months in prison. Compare that to the people before.' And the amnesty? Does he still regret accepting it? 'Actually now I think it was a nice decision, because there is war with Ukraine so nobody cares about Greenpeace any more. We could easily have got three years in prison and nobody would care.'

Popov got promoted, Kieron got married, Dima quit smoking and Pete went home to Maggy. From now on Pete Willcox is going to be more careful about which direct action protests he signs up for. 'I got a little too close there in Russia.' But he's here in Rotterdam, on the bridge of the *Rainbow Warrior*. 'I'm still campaigning to change our thirst for fossil fuels. Turning over the world to the oil and coal companies is not an option.'

Dima's great-grandfather helped build and sustain an empire, his father and grandfather were leading lights in a movement that brought it down, but because of a protest that lasted a few minutes it's likely Dima will never again see the country of his birth. And now, despite everything he went through out there, Gazprom is bringing that Arctic oil into Europe.

'I went to the Arctic to actually do something,' he says, 'so there's nothing to regret. Scientists say we can only stay safe if we keep atmospheric carbon dioxide levels below about 350 parts per million. Three hundred and fifty, that's the limit. We can't stay above that for long without fucking it all up. If it hits 400 then our kids are gonna be in very big trouble indeed. When my great-grandfather was jailed, CO_2 stood at 290. When my grandfather was jailed it was 310. When my father was jailed it was at 320. Three months before we took action at the *Prirazlomnaya*, NASA announced that atmospheric concentrations of carbon dioxide had just hit 400. It's happening. This is not a rehearsal. This requires resistance.'

Pavel Litvinov sees similarities and differences between his iconic protest in 1968 and the stand taken by his son half a century later. He faced certain trial and jail or exile, whereas his son didn't think he would face the full wrath of the Kremlin. 'The protests are different, of course,' says Pavel, 'but in a way the challenge that Dima and I both had in front of us was similar. We both wanted to speak up for somebody who was attacked by a large totalitarian government. In my case we spoke in defence of a small country, Czechoslovakia, which was suddenly oppressed by its big neighbour. And in the case of Dima he was speaking for the Arctic, which also didn't have its own defence, and to some degree the defence of the Arctic is a metaphor for the defence of humans and human rights. It is our life, because if the Arctic cannot survive then neither can we. It is our canary in the coal mine. If life is unbearable there then it will become unbearable to us. So there was a similarity. You try to raise a voice because you have nothing but your voice. A voice to speak up for something which cannot speak up for itself.'

In Murmansk SIZO-1 the prisoners are still fighting the regime, but they're not winning. 'I got a letter just two weeks ago from my cellmate,' says Denis. 'The situation there is terrible now, it's much worse than in our time. Those conditions were good, created specially for us. We had two or three cellmates in one room and we were allowed this *doroga*. We had TV sets and some people from a local NGO came to check on us. But when we left the prison nobody came to them and the authorities took away their TV sets. Now there are body searches and night searches almost every day and the meals have become water.'

Martin Sixsmith – the ex-BBC man – is on his way to lunch with the former director of MI5, when he makes time for a cup of Darjeeling tea with some of the campaigners. Together they conduct a post-mortem on why Putin freed the Arctic 30.

'One word,' says Sixsmith. '"Sochi". Putin was clearing the decks before the Olympics. It wasn't a surprise that Pussy Riot got out, and I wasn't at all surprised that the Arctic 30 got out. Khodorkovsky, that was the big surprise. It's the old thing about speaking quietly while carrying a big stick. Greenpeace had that stick. The Olympics.

'You guys said your plan was to give him "a wide turning circle". That was sensible. It gave him room to back down. Right now our governments are hammering away at Putin, accusing him of everything without understanding his dilemma, giving him no margin to make concessions and be sensible. It was absolutely clear that Greenpeace had to give him that opportunity to back down without losing face. He didn't want to keep the thirty in jail for ever, they would have been a thorn in his side. So he wanted to let them go, but he wanted to do it on his terms. It was the right thing to not hammer away at him, shouting and screaming. That way he could present the release as an act of magnanimity rather than him being bullied into it.

'If you look at it objectively, I think Putin played it just right. He showed himself to be tough then he let the thirty go. He showed himself to be magnanimous, having made his point, so everybody was happy. Yes, Greenpeace was petrified their people would never get out, but they were happy eventually. But most importantly his voters were extremely happy because they saw him standing up for them. Putin's image is this non-drinker, a judo fanatic, ex-KGB, takes no nonsense, dresses smart and stands up to the West. Stands up to people like the activists on that ship. So it was really important for his image to do what he did. When Putin does his analysis he'll probably think he came out on the plus side.

'Was it naïve of Greenpeace to think they could go in there, poke the bear and walk away? I assume the campaign leaders took all that into account when they went out there. They knew they'd be arrested, and in terms of publicity, having the guys arrested and jailed was a PR bonanza. It was unfortunate for the ones in jail, but good for the campaign.'

By the time they were freed, 2.7 million people had called for the release of the Arctic 30. Millions more are demanding a sanctuary at the top of the world where oil drilling and industrial fishing are banned. Something similar already exists in Antarctica after a campaign that took nearly twenty years to win. The push for an Arctic sanctuary may take longer, but the movement is mobilising. The fossil fuel companies have colonised almost every corner of the Earth, but if that movement can draw a line in the ice, if it can make its stand in the Arctic and win, then it can roll south and challenge the rule of oil across the globe.

Two weeks before the action at the *Prirazlomnaya*, one of the seven Arctic states, Finland – Sini's home country – became the first nation to join the call for an Arctic sanctuary. Six months later the European Parliament echoed that call.

Sometimes someone just has to jump first.

'As long as they continue with their dangerous plans then we're going to be there,' says Sini. 'The Arctic oil industry has decided to keep going, so we have to keep going too. It's not like we want to, but standing against them gives me a belief that we can actually win.'

Right now Sini's boat is drifting on the wake, floating away from the jetty as the *Mikhail Ulyanov* comes in. It's nearly docked now. She can see the faces of the Russian crew leaning over the railings and staring down at her. If she's going to jump, it has to be now.

She throws herself forward, for a split second she's hanging in the air then she crashes into the water. The cold is paralysing, she sinks below the surface but her life jacket lifts her, she gasps for air, shakes her head then kicks her legs. Back on the RHIB, Phil rolls his eyes. Everything inside him is saying, *oh shit, now you have to jump as well*. He hates swimming and he's got all this kit strapped to him. 'But I can see the ship, that big bastard ship, it's so close and if we can get to the jetty we stand a real chance of stopping it. And Sini was already swimming for it.' He coughs into his hand and steps up onto the side of the RHIB. 'And I just did it. I threw myself in.'

The police take a moment to notice what's happening, but the cops on the jetty don't move and the boats in the water are too far away to reach them. Sini turns her head and looks back and sees Phil swimming behind her. And the other climbers are launching themselves into the sea as well. A moment later they're all in the water, kicking hard, six of them, all weighed down with kit but getting closer. Then Sini reaches out and grabs the bottom rung of the ladder and hauls herself up.

Faiza is watching from the deck of the *Argus,* but the huge hull of the *Mikhail Ulyanov* is blocking her view and she can't see what's happening. Then her phone rings. It's the team on the jetty. They tell her they're hanging from the ladder, stopping that tanker from unloading its cargo of Arctic oil.

'For me it was very logical,' Sini remembers. 'The issue hadn't changed, it was the same fucking dirty oil, why wouldn't I protest against it? They'd started drilling and this was the first oil coming from that platform. My motivations hadn't changed in jail, if anything they'd become stronger. So in the end, whether or not I'd jump, I guess it wasn't really a question.'

AUTHOR'S NOTE

In September 2013 I took a phone call from Mads Christensen, at the end of which he asked me to lead the international media team pushing for the release of the Arctic 30.

'We need to make them famous,' he said. Then he hung up.

I lowered the phone and wondered if I'd actually agreed to take the job. I didn't say I wouldn't do it, so I guessed that meant I was doing it. Suffering as I do from acute imposter syndrome, I thought momentarily about calling him back and politely declining the offer, but I knew half the people in jail and some of them were good friends. I'd climbed power station chimneys with them and broken into polluting factories at their side. I'd sailed to Greenland with Iain Rogers, Colin Russell and Mannes Ubels. I was slated to join the *Sunrise* on its mission to the Russian Arctic, but my boss wouldn't let me go and Alex went instead. It could have been me in that Russian prison cell.

The next three months were spent in a maelstrom of black coffee, boiled sweets and fear, as we managed a publicity operation spanning dozens of countries across the world. I neglected the people I loved and gave myself over entirely to the Room of Doom. The only times I left that place for an extended period were a strategy meeting in Copenhagen, a failed attempt to spend a weekend away with my girlfriend, and the moment the Arctic 30 were released.

After the news from Ana Paula's hearing I sped to St Petersburg and checked into the Peterville hotel. Half an hour after Sini and Camila were freed I stood at the end of a corridor watching them bouncing on their heels and grinning wildly as they opened the door to their room. Behind them Ana Paula and Anne Mie were hugging each other. Since September those faces had stared back at us from posters on the wall of the bunker and from newspaper front pages, and now they were there, in front of me. Perhaps it was a product of the pedestal we all put them on while we were working for their release, but I remember thinking they all looked a lot *shorter* than I imagined they'd be.

The following evening I sat with Frank in the Peterville café, each of us nursing a beer, and he told me about that night at SIZO-1 when his cellmate yanked the U-bend from the wall and he heard the voice of Roman Dolgov broadcasting through the plumbing system. Then Frank told me more stories – about Popov, his cellmates, the guard who loved Depeche Mode – and as he spoke I jotted down notes on a napkin. I thought those stories might be the basis of an email to my colleagues around the world to whom I was sending updates from St Petersburg. Frank and then Anthony reeled off more tales, I scribbled on more napkins and stuffed them into my pockets, but eventually I stopped writing and just sat back and listened. And at some point that night I thought somebody should write a book about it all.

Back then I was still utterly consumed with the campaign to get them home and, more immediately, the job of managing the huge media interest in the men and women pouring out of jail. Dozens of journalists were travelling to St Petersburg or had already arrived. A Swiss TV crew refused to leave the hotel bar until they'd interviewed Kruso. A British newspaper reporter was stalking the corridors looking for Alex, determined to negotiate an exclusive deal.

Five weeks after they were freed, I came home with the British activists, and soon after they stepped onto the concourse at St Pancras station I slipped away, took a train home to my family for a belated Christmas and collapsed exhausted onto my bed. For two days I didn't leave the house. But when I did surface, I pulled those napkins from various pockets and read through them again, and I thought, yes, somebody should definitely write a book about it all.

I left it a month, then started interviewing the activists, in person and on Skype. And I spoke to many of my colleagues from the hubs in Copenhagen, London, Amsterdam and Moscow. I was fortunate that a team of volunteers transcribed those interviews – more than forty hours in total (you can listen to some of them at www.donttrustdontfeardontbeg. com). I then retreated to a house in the countryside for a fortnight, carrying a three-inch pile of printed interviews and a pack of luminous magic marker pens.

My plan was to highlight the quotes that might conceivably be moulded into a narrative, but by the time I reached the end of the pile, most of the sheets were almost entirely covered in bright green, pink and yellow ink. So many stories, so many characters. It felt like there were a dozen books in there, and I was yet to find a single one of them.

Only by concentrating on four or five of the thirty –
and a smaller number of the campaigners – did something
digestible begin to emerge. The result is a book that fails to tell
a host of remarkable stories. An entire second volume could
be written about the experiences of Faiza, Kieron, Roman,
Colin, Andrey, Denis, Anthony and the others.

I had a few stories myself, and in the first draft I sought to
tell them. The text was littered with 'I', 'we' and 'us'. But when
I read back through the chapters, those parts – the ones told
in the first person – struck a bum note, like a chord played
on a badly tuned piano. This isn't my story; it's the story of
the activists who were jailed in Russia for scaling an Arctic oil
platform. Trying to relate my own experiences amid those tales
from SIZO-1 felt like jumping out of the crowd at a football
match and running onto the pitch dressed in a replica kit.

So I exorcised myself from this book. Nevertheless, some of
the events relayed here are ones I witnessed or participated in.

That three-inch pile of paper contained some gaps of
recollection and detail. Dima's memory for conversations,
especially between him and Popov, was extraordinary, but
not everyone remembered so easily. Therefore I occasionally
reconstructed details and dialogue before checking my efforts
with the activists to ensure accuracy.

The English of some of the Russian prisoners was so
limited that the articulation of a single sentence would take
an age, and they often used sign language as much as the
spoken word. I have tried to reflect that in the nature of their
dialogue, whereas the words spoken in Russian – for example
by Vitaly to Dima – are more immediate and expressive
because they were said in the mother tongue.

Vitaly, I should say, is not his real name. I have changed the names and some identifying characteristics of the Russian prisoners because I was not in a position to ask their permission to recount their stories. I feared – perhaps unrealistically, but who can say? – that those still behind bars might face retribution for some of the things I report them saying and doing, not least the support they offered their Greenpeace cellmates. I would not like to make the job of Popov any easier.

I have also altered identifying characteristics of some of the other Russians featured in this story, again to minimise the possibility of retribution being wrought. The scene in which Phil smuggles the footage out of prison was shifted from the place it actually happened, to protect 'Mona' from serious criminal charges (it's best she doesn't find herself in the women's section of SIZO-1).

I wanted to give a more realistic portrayal of the incarcerated women than the one offered by some media outlets during their time in jail. In many newspapers there was an assumption that the women would be coping less well with their ordeal than the men. Alex became keenly aware of the discrepancy when she googled her name at the Peterville hotel. Many of the photographs were of her crying in the cage at her first appeal in Murmansk. That moment appeared to define her, and it is something for which Greenpeace – myself included – bear some responsibility. In some countries the organisation made Alex the face of the campaign and used the image of her at the appeal hearing on advertisements and leaflets. When, after her release, she saw how her tears had been used, she tried but failed to hide her disappointment.

I tried but failed to hide my culpability. I had authorised those adverts.

'These stereotypes piss me off,' she told me. 'I don't like the way everyone portrays the women compared to the men. When I needed to cry, I cried. When I needed to scream, I screamed. Being true to yourself and your emotions isn't a weakness, it's not something I'm ashamed of. But the way the articles are written, people were more worried about me because I cried, or they were more worried about the women because we're women.'

In reality the women were as strong – if not stronger – than the men. Often the younger activists coped better than the older ones, in part no doubt because they were less likely to have long-term partners or children back home. In St Petersburg after their release the women were a source of constant, unbridled, positive energy. I remember one night standing in the Helsinki bar, drinking shots of vodka with Kieron, watching Alex, Camila, Sini and Faiza dragging everyone – Russian strangers included – off their chairs and onto the dancefloor. A Michael Jackson track had just come on, one that used to be played on Bridge TV back at SIZO-1, and they were recreating the dance moves they made in their cells. And I remember saying to Kieron, 'When I grow up I want to be Sini Saarela.'

Sometime later I asked Kieron how he managed to survive each night back in that isolation jail in the Russian Arctic.

'You don't really,' he said. 'You just tell yourself that the fifteen years might happen. You tell yourself to get used to that sick feeling because eventually that might be reality. I wrote a letter to Nancy' – his girlfriend, later his wife – 'and I said to her, look, if this goes really bad there's no way I expect you to wait for me. My big fear was missing out on having a family because being forty-four in fifteen years I would've lost the person I wanted to marry, lost my career, the people I love, and every semblance of joy and success I've had would be totally obliterated by this fifteen years of darkness. I actually remember thinking I won't make it fifteen years. I honestly didn't think I'd get that far. Would I top myself? Yeah, I think I probably would. But you have to keep hoping.'

As hard as it is to imagine, many of the prisoners' families had it just as bad, and I regret not being able to tell more of this story from their perspective. I interviewed some of them but the narrative became increasingly cluttered, so instead I sought to have Pavel Litvinov represent them in this account. Despite his remarkable history, he was, in late 2013, just one more terrified parent. In the absence of more of their experiences, this book is dedicated to the families of the Arctic 30.

It was an incredible honour for me to interview Pavel. I knew of his iconic protest in Red Square, having long ago read *Lenin's Tomb* – David Remnick's remarkable book about the fall of the Soviet Union, in which he features prominently. But despite being moved by his courage and inspired by his activism, I didn't realise until September 2013 that he is Dima's father. Only when he brought his considerable energy to the campaign to free his son did I become aware that this was the same Pavel Litvinov.

I sought to raise an important question in the book, but it's not one I tried myself to answer. Was Greenpeace naïve to scale the side of a Russian Arctic oil platform and expect to sail away unmolested by Putin's judicial machine? I have heard a dozen different answers. One senior colleague told me he thought everything had gone exactly according to plan, that creative disruption brings about change, that if we're serious about challenging the fossil fuel behemoths then people are going to have to go to jail. Equally I have heard many people at Greenpeace claim that nobody could have predicted what would happen. After all, a year earlier the crew of the *Arctic Sunrise* had launched an almost identical protest and the FSB had done nothing. And others – including some of the thirty – say yes, it was naïve of Greenpeace to have sailed for the *Prirazlomnaya* and not thought jail for its activists was the likely outcome.

In a sense the question is subjective. There is perhaps no right or wrong answer. A more relevant question might be, 'Was it worth it?' And that is for nobody to answer but the thirty themselves. In interviews in St Petersburg most of them said yes, it was. But some would disagree, not least because of the impact on their families, and who could blame them?

Another question often asked is, 'Did it change anything?' I would say it did. Securing action on climate change is hard. We are all implicated in the fossil fuel economy, and most political and corporate leaders are yet to determine that it's in their interests to truly roll back dependence on oil, coal and gas. We are caught in the same paradox as Dima was in his cell at SIZO-1 when he started smoking then realised he'd gain nothing by giving up. It only took one prisoner to light up to make it pointless for the others to quit, because they all shared the same air. And it's a bit like that with climate change. Those with the power to effect change – prime ministers, presidents and CEOs – think there's nothing in it for them to act alone, that it's only worthwhile quitting fossil fuels when everybody else does, because we all share the same atmosphere.

The campaign to save the Arctic is not yet won. In fact it's barely begun, and we don't have much time left. But direct action – the manifestation of resistance, the ignition of controversy from apathy – speeds up the national and global conversation, it short-circuits ponderous political cultures and forces uncomfortable issues onto the agenda. It is asymmetric campaigning, a leveller, a way for the weak to take on and beat the strong, to make governments and corporations act against their immediate interests. To break that paradox.

The night he left Russia for home, I sat with Dima in the hotel bar as he wrote his statement for the press. With my laptop balanced on his knees he bashed away at the keyboard, then he passed me the computer and said, 'Release that.' He got up and walked towards the door, his pink bag slung over his shoulder, his wife's hand in his, and I looked down at what he'd written.

I've never regretted what I did, not once, not in prison and definitely not now. Sometimes you just have to stand up and ask to be counted, and that's what we did in the Arctic. They didn't throw us in jail for what we did, they locked us up because of what we stood for. The Arctic oil companies are scared of dissent, and they should be. They may have celebrated when our ship was seized, but our imprisonment has been a disaster for them. The movement to save the Arctic is marching now. Our freedom is the start of something, not the end. This is only the beginning.

ACKNOWLEDGEMENTS

My apologies to the campaigners whose efforts are not recognised in this book. There are too many to name here, but it's a particular regret that I couldn't include the exploits of Jessica Wilson, Elena Polisano, Sol Gosetti and Dima Sharomov. The contributions of Nora Christensen, Ben Ayliffe, Fabien Rondal, Neil Hamilton, James Turner, Rachel Murray, Iris Andrews and Jasper Teulings were more significant than the brevity of their mentions suggests. The same can be said for the heroic staff of Greenpeace Russia. I thank them for their understanding.

Thanks also to my agent, David Godwin, for steering a goldfish through a shark pool, and to Laura Hassan and Carl Bromley for their advice and support. I owe a great deal to the friends who took time to read early drafts before proffering invaluable advice, and to John Sauven and Robin Oakley at Greenpeace for supporting me from an early stage. Very special thanks to the volunteers who transcribed interviews, and of course to the interviewees themselves – especially to the released prisoners who gave me their time.

While the thirty were in jail I barely spent any time at home, and when I was there I lay awake on the bed reading tomes of Russian political analysis. My girlfriend wondered if we might go away for a while. I told her I couldn't possibly leave the campaign at such a critical juncture. She said I'd been saying that for weeks, so we compromised and booked a weekend in France. At a café table in the shadow of the Abbaye de Saint-Riquier she sat opposite me, and not for the first time she sipped coffee after coffee in silence as I sat over my laptop with earphones clamped to my head, a microphone at my lips, looking like a pretend helicopter pilot, fielding Skype calls from around the world.

Lorna, I'm sorry about that. Thank you for your love and constant encouragement.

ENDNOTES

1 http://world.time.com/2013/03/06/the-magnitsky-trial-russia-places-a-dead-man-in-the-dock/

2 http://www.oedigital.com/component/k2/item/5305-gazprom-makes-arctic-advances

3 http://www.greenpeace.org/usa/en/campaigns/ships/the-rainbow-warrior/20th-anniversary/the-terrorist-plot/

4 http://www.mensjournal.com/magazine/pete-willcox-high-seas-avenger-20140324

5 http://www.greenpeace.org/international/en/news/Blogs/makingwaves/safety-pod/blog/46713/

6 http://www.bbc.co.uk/news/world-europe-24879073

7 https://docs.google.com/file/d/0B_VQeHLcziV_VkkodlBSRTQ2Yjg/view?sle=true

8 http://www.theguardian.com/environment/2013/sep/25/vladimir-putin-greenpeace-kenya-mall-attack

9 http://www.greenpeace.org/international/Global/international/photos/climate/2013/GP04WD8.jpg

10 Simon Sebag Montefiore, *Young Stalin* (London, 2007), pp. 182–3.

11 Lee B. Croft, Ashleigh Albrecht, Emily Cluff, Eric Resmer, *The Ambassadors: U.S.-to-Russia/Russia-to-U.S.* (Phoenix, 2010), p. 150.

12 Helen Rappaport, *Conspirator: Lenin in Exile, The Making of a Revolutionary* (London, 2010), pp. 136–7.

13 http://query.nytimes.com/gst/abstract.html?res=9B0CEFD9173EE233A25751C0A9649C946997D6CF

14 Croft *et al.*, p. 150.

15 http://www.britannica.com/EBchecked/topic/344401/Maksim-Maksimovich-Litvinov

16 http://books.google.co.uk/books/about/Maxim_Litvinov.
html?id=wCEhkgAACAAJ&redir_esc=y

17 http://content.time.com/time/covers/0,16641,19330424,00.html

18 Jeffrey Herf, *The Jewish Enemy: Nazi Propaganda During World War II and the Holocaust* (Harvard, 2006), p 97.

19 Aleksandr Moiseevich Nekrich, Adam Bruno Alam, Gregory L. Freeze, *Pariahs, Partners, Predators: German–Soviet Relations, 1922–1941* (Columbia, 1997), p. 109.

20 Nekrich *et al.*, p. 109.

21 Roderick Stackelberg, Sally A. Winkle (eds), *The Nazi Sourcebook: An Anthology of Texts* (London, 2002), p. 245.

22 Victor Israelyan, *On the Battlefields of the Cold War: A Soviet Ambassador's Confession* (Pennsylvania, 2003), p. 10.

23 Jonathan Haslam, *Russia's Cold War: From the October Revolution to the Fall of the Wall* (Yale, 2011), p. 75.

24 V. M. Molotov and Feliz Chue, *Molotov Remembers: Inside Kremlin Politics* (Chicago, 1993), pp. 68–9.

25 *Young Stalin*, p. 183.

26 http://www.nytimes.com/1997/06/20/world/lev-kopelev-soviet-writer-in-prison-10-years-dies-at-85.html

27 http://www.nytimes.com/1997/06/20/world/lev-kopelev-soviet-writer-in-prison-10-years-dies-at-85.html

28 http://www.nytimes.com/1997/06/20/world/lev-kopelev-soviet-writer-in-prison-10-years-dies-at-85.html

29 http://ausstellung-gulag.org/en/476/

30 http://www.nytimes.com/1997/06/20/world/lev-kopelev-soviet-writer-in-prison-10-years-dies-at-85.html

31 Stephen F. Cohen, *The Victims Return: Survivors of the Gulag After Stalin* (New York, 2011), p. 16.

32 http://www.independent.co.uk/news/people/obituary-lev-kopelev-1256877.html

33 http://ausstellung-gulag.org/en/476/

34 Gale Reference Team, Biography – Litvinov, Pavel (1940–), *Contemporary Authors* – 2002 (digital download)

35 David Remnick, *Lenin's Tomb: The Last Days of the Soviet Empire* (New York, 1993), p.17.

36 Pavel Litvinov, *Dear Comrade: Pavel Litvinov and the Voices of Soviet Citizens in Dissent* (1976)

37 *Dear Comrade*

38 *Lenin's Tomb*, pp. 18–19.

39 *Lenin's Tomb*, p. 19.

40 *Lenin's Tomb*, p. 19–20.

41 Yuily Kim song lyrics http://www.bards.ru/archives/part.php?id=6188

42 http://www.greenpeace.org/international/en/press/releases/
    Emergency-solidarity-protests-worldwide-to-free-journalists-and-
    Greenpeace-activists-held-in-Russian-prison/

43 http://rt.com/business/2012-sponsorship-football-marketing-882/

44 http://www.whoi.edu/oil/deepwater-horizon

45 P.J. v Brandvik, K.R. Sørheim, I. Singsaas and M. Reed 'Short
    State-of-the-Art Report on Oil Spills in Ice-Infested Waters: Oil
    Behaviour and Response Options', *SINTEF*, 19 May 2006.

46 http://ocean.si.edu/gulf-oil-spill

47 http://www.reuters.com/article/2014/05/13/
    energy-arctic-idUSL6N0NZ3GV20140513

48 http://news.bbc.co.uk/1/hi/programmes/newsnight/9483790.stm

49 http://www.greenpeace.org/australia/en/news/climate/
    In-30-years-weve-lost-75-of-the-Arctic-sea-ice/

50 http://thinkprogress.org/climate/2013/02/14/1594211/
    death-spiral-bombshell-cryosat-2-confirms-arctic-sea-ice-volume-
    has-collapsed/

51 http://www.noaanews.noaa.gov/stories2013/20130412_arcticseaice.
    html

52 http://uk.reuters.com/article/2011/09/13/uk-cairnenergy-
    idUKTRE78C1D420110913

53 http://www.theguardian.com/environment/2013/jan/18/
    shell-oil-drilling-arctic-environment

54 http://www.bp.com/en/global/corporate/gulf-of-mexico-
    restoration/deepwater-horizon-accident-and-response/completing-
    the-response.html

55 http://www.greenpeace.org.uk/blog/climate/why-shells-
    spill-response-plan-dogs-breakfast-20120221

56 http://barentsobserver.com/en/sections/articles/
    no-way-clean-oil-spill-under-ice-canadian-expert

57 http://online.wsj.com/news/articles/SB10001424127887323320404
    578215041848706404

58 http://www.adn.com/article/emails-say-shell-containment-dome-crushed-beer-can-test

59 http://www.telegraph.co.uk/finance/newsbysector/energy/oilandgas/9712687/Shell-Alaska-boss-There-will-be-spills.html

60 http://www.theguardian.com/business/2014/apr/04/shell-oil-safety-warnings-moved-drill-ship-us-coast-guard

61 http://www.ft.com/cms/s/0/486a34d4-898a-11e3-abc4-00144feab7de.html#slideo

62 http://www.greenpeace.org/international/en/news/Blogs/makingwaves/10-reasons-to-take-action-to-stop-gazproms-pr/blog/46766/

63 http://www.greenpeace.org/international/en/news/Blogs/makingwaves/10-reasons-to-take-action-to-stop-gazproms-pr/blog/46766/

64 http://shelf-neft.gazprom.ru/en/?type=larn

65 http://www.greenpeace.org/international/en/news/Blogs/makingwaves/10-reasons-to-take-action-to-stop-gazproms-pr/blog/46766/

66 Martin Sixsmith, *Russia: A 1,000-year Chronicle of the Wild East* (London, 2011), p. 510.

67 *Russia*, p. 511.

68 *Russia*, p. 511.

69 *Russia*, p. 511.

70 Charles Emmerson, *The Future History of the Arctic* (London, 2010), pp. 227–8.

71 *Russia*, p. 511.

72 *Russia*, p. 509.

73 http://pages.uoregon.edu/kimball/Putin.htm

74 http://www.globalsecurity.org/military/world/russia/energy.htm

75 http://www.independent.co.uk/environment/nature/countries-lay-claim-to-arctic-in-battle-for-oil-and-gas-reserves-2087040.html

76 http://origins.osu.edu/article/russia-and-race-arctic

77 http://www.amnesty.org.uk/russia-crackdown-human-rights-lgbt-gay-law-protest-censorship-pussy-riot-sochi#.VCJuvFb4vwI

78 http://segelreporter.com/wp-content/uploads/2013/10/193757266-4e69c8ea-0547-4f60-a28a-ae07a21ea275.jpg

79 http://www.greenpeace.org/international/en/news/Blogs/
    makingwaves/the-mothers-of-the-disappeared-want-the-arcti/
    blog/47004/

80 http://www.greenpeace.org/africa/en/News/news/
    Desmond-Tutu-joins-the-call-to-free-the-Arctic-30/

81 http://www.greenpeace.org/international/en/press/releases/
    Eleven-Nobel-Peace-Prize-winners-write-to-Russian-President-
    Vladimir-Putin-over-Greenpeace-case/

82 https://www.youtube.com/watch?v=8xRT7wQiebw

83 http://www.reuters.com/article/2013/10/17/
    russia-greenpeace-gazprom-idUSL6N0I73BV20131017

84 http://www.mensjournal.com/magazine/
    pete-willcox-high-seas-avenger-20140324

85 http://www.mensjournal.com/magazine/
    pete-willcox-high-seas-avenger-20140324

86 http://www.mensjournal.com/magazine/
    pete-willcox-high-seas-avenger-20140324

87 http://www.mensjournal.com/magazine/
    pete-willcox-high-seas-avenger-20140324

88 http://www.crmvet.org/tim/timhis63.htm#1963selma1

89 http://www.crmvet.org/info/lithome.htm

90 http://www.nps.gov/nr/travel/civilrights/al4.htm

91 http://www.nps.gov/semo/historyculture/index.htm

92 http://news.bbc.co.uk/onthisday/hi/dates/stories/march/28/
    newsid_4264000/4264241.stm

93 http://iipdigital.usembassy.gov/st/english/publication/2009/01/
    20090107151130srenod0.5167658.html#axzz3EDtsZjvF

94 http://www.mensjournal.com/magazine/
    pete-willcox-high-seas-avenger-20140324

95 http://www.sciencephoto.com/media/152694/view

96 http://www.mensjournal.com/magazine/
    pete-willcox-high-seas-avenger-20140324

97 http://www.mensjournal.com/magazine/
    pete-willcox-high-seas-avenger-20140324

98 http://www.academia.edu/1005097/
    The_Rainbow_Warrior_bombers_media_and_the_judiciary_
    Public_interest_v_privacy

99 http://www.theguardian.com/environment/2005/jul/15/activists.g2

100 http://www.academia.edu/1005097/The_Rainbow_Warrior_
    bombers_media_and_the_judiciary_Public_interest_v_privacy

101 David Lange, *My Life* (Auckland, 2005), pp. 222–3, pp. 274–5.

102 Terry Crowdy, *Military Misdemeanours: Corruption, Incompetence,
    Lust and Downright Stupidity* (Oxford, 2007), p. 246.

103 http://www.thetimes.co.uk/tto/news/world/article1980551.ece

104 http://www.theguardian.com/environment/2007/may/25/usnews.
    france

105 http://www.mensjournal.com/magazine/
    pete-willcox-high-seas-avenger-20140324

106 *Dear Comrade*

107 http://www.rferl.org/content/article/1059107.html

108 http://www.bloomberg.com/news/2013-11-06/dutch-urge-release-
    of-greenpeace-crew-in-court-clash-with-russia.html

109 http://www.dailymail.co.uk/debate/article-2458302/DOMINIC-
    LAWSON-Putins-brute-Greenpeace-bigger-menace-future.html

110 http://www.thesundaytimes.co.uk/sto/news/article1323429.ece

111 http://best-museums.com/russia/85-museum-crosses.html

112 http://www.poetryfoundation.org/bio/anna-akhmatova

113 http://www.spb.aif.ru/society/135936

114 http://www.spb.aif.ru/society/135936

115 http://articles.latimes.com/1999/oct/17/news/mn-23277

116 http://www.greenpeace.org/international/en/press/releases/
    Greenpeace-Current-draft-of-Russian-amnesty-does-not-include-
    Arctic-30/

117 http://www.greenpeace.org/international/en/press/releases/
    Russian-parliament-votes-for-amnesty-for-Arctic-30/

118 http://www.telegraph.co.uk/news/worldnews/vladimir-
    putin/10527779/Vladimir-Putin-pardons-oil-tycoon-
    Mikhail-Khodorkovsky-in-Amnesty.html

119 http://www.theguardian.com/world/2013/dec/19/
    russia-never-worked-edward-snowden-nsa-putin

120 http://eng.kremlin.ru/news/7040

121 http://www.ship-technology.com/projects/mikhail_ulyanov/

INDEX